Land Records Management in India

About the Editors

Dr Amir Afaque Ahmad Faizi an officer of Bihar Administrative Service, presently working with the Academy as Professor Center for Rural Studies. Dr Faizi has served Ministry of Rural Development, Government of India, as Addl P.S. to Minister for Rural Development, Dr Raghuvansh Prasad Singh, and as Consultant, NMMU, NRLM later on. He did three Master programmes, one in Public Administration, one in Rural Studies in which he was awarded Gold Medal by His Excellency APJ Abul Kalam the President of India, and in Governance and Development from IDS, University of Sussex, UK. He got his doctorate from K.R. Narayanan Center for Dalit and Minorities Studies, Jamia Millia Islamia on Self Help Group and its impact on marginalized communities. Many of his work has been published in national and international journals of Public Administration and Rural Development.

Hari Charan Behera, PhD, is presently a Faculty Member of the Indian Statistical Institute (ISI) Giridih. Before joining the ISI, he was with the Centre of Rural Studies, LBSNAA as a capacity of Assistant Professor. He has interest in the field of land governance, agrarian studies and rural development.

Land Records Management in India

A Plea for Reforms

Edited by
A.A.A. Faizi
Hari Charan Behera

Published for
LBS National Academy of Administration
Mussoorie

by

CONCEPT PUBLISHING COMPANY PVT. LTD.
NEW DELHI-110059

ISBN-13: 978-93-5125-095-1

First Published 2014

Published and Printed by

Concept Since 1974

Concept Publishing Company Pvt. Ltd.
A/15-16, Commercial Block, Mohan Garden
New Delhi-110059 (India)
T: 25351460, 25351794, *F*: 091-11-25357109
E: publishing@conceptpub.com, W: www.conceptpub.com
Editorial Office: H-13, Bali Nagar, New Delhi-110 015, India.

Cataloging in Publication Data--*Courtesy:* D.K. Agencies (P) Ltd. <docinfo@dkagencies.com>

Land records management in India : a plea for reforms / edited by A.A.A. Faizi, Hari Charan Behera.
p. cm.
Includes bibliographical references and index.
ISBN 9789351250951

1. Land titles--Registration and transfer--Technological innovations--India. I. Faizi, Amir Afaque Ahmad. II. Behera, Hari Charan, 1978- III. Lal Bahadur Shastri National Academy of Administration.

DDC 346.540438 23

Foreword

Land, apart from being a means for survival, is also fundamental for economic growth and prosperity. Every nation in the world has a fundamental structure for land management, so as in India. In India we have inherited a structure of land administration which is more or less similar with that of the colonial India. Amidst mountainous pressure on land rising over the period due to change in nature of land-use pattern, reform process in the country is rather slow and sluggish. In certain areas like ICT application in land records management, India is at par with most advanced nations in the world. Remote sensing technology in India is amongst the best in the world. But unfortunately its applications in land records management are not wide enough. But our institutional and legal reforms have not been compatible with that of best technology available for land records management. The information pertaining to land, land ownership, land tenure, crop updation, etc., are not updated regularly. The village maps or land parcel maps in many instances require updation through revisional survey. In some States, particularly in majority areas of north-eastern states, fresh cadastral survey is required. Therefore, reform process is must to upgrade the management process.

The National Land Records Modernization Programme (NLRMP) is a step forward to approach new methodology for cadastral survey, property registration, and maintenance of land records. It is an amalgamation of multiple methodologies, principles and guidelines. But it requires committed effort for implementation and provides scope for guaranteeing titling system in the country.

I am glad that the Centre for Rural Studies is doing sincere effort to collate ideas, thoughts and innovations on this issue and bringing out a book which can be best useful for the Centre as well as several States to pursue ideas into action.

Padamvir Singh, *IAS*
Director, LBSNAA

Preface

A good land administration boosts good governance to deliver transparent, accountable, efficient and quality service to the citizen as well as other stakeholders. In the backdrop of recent economic trend, reforms in land administration and land records management has become a focal point for planners, policy makers and development architects. In India, the issues pertaining to land records are complex. Land record is a state subject. Different states have different practices in the area of survey and settlement operation, creation, updation, and maintenance of land records, which are often guided by state revenue codes and guidelines. Departments namely, survey and settlement department, the registration department and the revenue or land records department work in silos with little interaction among each other. The management of land reforms, land records management, survey and consolidation, the registration and mutation operations all too differ accordingly. The whole system of land records management involves huge cost, in terms of time spent, money spent and labour required. It involves huge manpower, complicated technology and methodologies. The transaction cost involved in land records management is high. We have the process of land ownership which is presumptive, but not conclusive one. Land related disputes are not less either. In some states the land records are in good shape, whereas in some other states they are not. Fresh survey and resurvey depends on the quality of land records already available in the states. Modern cadastral survey operation is required in different degree of passion and commitment. Both cost and technology requirements are all State specific. However, every State has to work in a particular direction to achieve larger vision of good governance in land administration through reforms. Considering existing issues such as poor service delivery, lack of transparency, accountability and inefficiency in the system of land records management, we are lagging far behind of many nations who have already best cadastre system and conclusive titling system.

I am glad that the Centre for Rural Studies is taking lead step to facilitate the Department of Land Resources at the Centre by organizing such workshops and bringing out publications. I hope the recommendations of the workshop will be executed in a definite time frame for larger benefit of the States and the citizens.

Prem Singh, *IAS*
Centre Director, CRS

Acknowledgements

We wish to express our sincere thanks and appreciation to Dr. Prem Singh, IAS, Centre Director, Centre for Rural Studies for his constant inspiration and encouragement to edit this book.

We owe our graditude to Shri B.B. Srivastava, IAS, (former) Secretary, DoLR and Shri Sanjeev Chopra, IAS, Joint Director, for their kind support to bring out this book.

Many people and Organizations have contributed at different stages in the completion of the book. Thanks are due to all the contributors of the book who are also the champions in the field of land record management.

This book covers maximum dimensions in Land Records Management of India. We indeed grateful to Ms. Nivedita P. Haran, IAS, Addl. Chief Secretary, Revenue and Disaster Management, Government of Kerala, for her immense support and contribution.

We also acknowledge Dr. Charanjit Singh, Director DoLR, MoRD, Government of India for his valuable contribution throughout the entire process.

Nonetheless, the cooperation of all the faculty and staff of CRS, particularly from Mr. Snehasis Mishra, Technical Consultant of CRS-LBSNAA, Mr. Ramesh Kothari, Mr. Deepak Kumar, Mr. S.S. Kharola, Mr. Dalip Bisht, Mr. Adesh Kumar, and Mr. Samar Kashyap is well appreciated. These persons have not only extended support but also their sincere effort on that topic make it easy.

Contents

Introduction

A.A.A. Faizi and *H.C. Behera*

Good governance through best practices in land administration has been a longstanding vision for developing countries. Major developing countries like India, Brazil, South Africa, Indonesia, and many others have tried to adopt a system of land administration that can be efficient, transparent and more effective in service delivery. While these countries pertain similar situation with high density population, frequent division of land parcels is quite so obvious. Frequent partitions of land due to transfer or inheritance of land, sale and purchase has led to creation of new records and modification of old records. Moreover, for development planning and execution land is fundamental which requires special attention from the key revenue functionary. Most development thinkers and planners believe that there is need for reforms in existing system of land administration and land records management which is still guided by colonial policy. In India, the existing system is more complex and diverse. The country inherits more or less similar practice of cadastral survey, creation and updation of land records even after six decades of independence where there is huge transition in demography, socio and economic policy. Land being a state subject, the system is further more complex and it is diverse in nature. The Centre only plays the role of guidance and facilitation. The Union Government creates and facilitates Acts, laws and legislations. It can direct the State but can't enforce its enactment, amend State laws and execution. The States pursue the State guidelines, amend laws and legislation but that gain carries no commonality in the system of land records management. Therefore, bringing commonality in structure of land policy and administration without affecting the sentiments of State, removing vested political interest, ideology and State sovereignty is a critical challenge.

What is existing system of Land Administration and Land Records Management?

Land question is the question of age (Pankhurst, 1883)[1]. The root of existing system of land records management is linked to colonial policy. The British realized that they must control land in order to stabilize their rule over the continent. Their first effort was to fix the legal owner of the land (Bandyopadhyay, 1993)[2]. Despite the fact that their primitive motive was to extract rent but that initiated the process of survey, creation and updation of land records, registration, etc., inherits more or less similar legislations and guidelines even today.

The management of both government and private lands remain same. While some guidelines are strictly followed but some other guidelines have not been easy to be implemented. For example, the revisional survey needs to be done in a gap of time period, but none of the States are able to catch up the momentum. The land records existing are either old or outdated. In most of the States there are old and outdated maps. In most cases the maps are torn and fragile. As a result, the States are recurring with the problems of land related disputes. The disastrous consequence is, however, the poor who are prone to losing land in the hands of powerful people with vested interest. The registration of document involves high transaction cost. The stamp fee is one of the highest in the world. But again there is no guarantee that the transaction can ensure foolproof measures. Mutation after registration takes long time and high transaction cost. The complexity of mutation process is one of the factors of huge backlog in the revenue department.

The other important factor is the administrative issues involved in land records management. The survey and settlement department, the land records division and registration department are three important administrative structures under land revenue administration. The survey and settlement division works for survey, measurement and creation of new land records that provides location, boundary detection, measurement, etc., which again in administrative term called cadastral survey. The survey and settlement department becomes responsible for creation of map of land parcel, called cadastral map, village boundary, etc. These survey records are managed by land record division. But the historic problem attached with the integration of cadastral maps with non-textual data or attributable data. Both spatial information and textual information should complement each other. In India, however, the maps are not renewed periodically, land parcels do not reflect accurate data. There are many developmental progress, new infrastructure, road, railway, building, etc., the original cartography can be totally renewed with immediate effect. Moreover, lands are partitioned, shared and redistributed. Many new land parcels are developed which are not updated in cadastral map. The revenue department has one set of information may be with or without proper updation, but for sure not proper updation in cadastral maps. Some way or the other there is lack of coordination between the departments. Moreover, the registration department does not ensure authenticity of information. The Indian Registration Act 1908 does not guarantee title of the land holder. There is no conclusive proof of the title of land holder. The registration process is not linked to automatic mutation. These three departments work in silos with very little dynamic interaction. The revenue codes and guidelines for land records management; which were introduced during British regime remain more or less same till date. Land is a State subject under Constitution of India. Thus, the Revenue Codes and guidelines vary across the States. With the evolution of technology; newer methods of survey and settlement; better methods of data keeping and retrieval; the need for revival of the system of land records management is actually felt. The synchronization of activities related to land records management among the three departments needs to be strengthened by leveraging technology.

The debate and deliberation on land records management is not new to India. The process of land records modernization started with the schemes of 'Strengthening Revenue Administration and Updating of Land Records (SRA and ULR) and Computerization of Land Records (CLR)' which began in 1987 and 1988 respectively. In 2008, Government of India launched another scheme called the 'National

Land Records Modernization Programme (NLRMP)' by merging two old schemes, viz., SRA and ULR and CLR. The main aims of National Land Records Modernization Programme are "to usher in a system of updated land records, automated and automatic mutation, integration between textual and spatial records, inter-connectivity between revenue and registration, to replace the present deeds registration and presumptive title system with that of conclusive titling with title guarantee". Apart from the National Land Records Modernization Programme, the Land Titling Bill and LA and RR Bill are some of the recent initiatives in this regard.

It is expected that the land related programmes can certainly enhance capacity of the multiple institutions dealing in various areas of land records management, survey and settlement and property registration. Further, the NLRMP has leveraged modern technology, led to creation of enabling environment through formulation of right policy for the States and Union Territories, brought about better coordination between Centre and State for effective implementation of the programme and led to capacity development at various levels of the revenue administration. However, there is a need for further reforms at various levels to execute and accelerate the process of modernization of land records.

The good governance in land administration is far more achievable without improvising the existing policy and execution process. Thus National Land Records Modernization Programme (NLRMP) provides plethora of scope for improvement in the existing practice, institutional renovation and proper land records management.

Objectives

1. To examine current trend and practices in land records management:
   (a) The current trends and practices in land records management and challenges in land administration; and
   (b) Land records management in tribal and other backward areas.
2. To look at technology Intervention in the area of cadastral survey, settlement and land records management:
   (a) Application of modern technologies, such as GIS, GPS,

ETS and other initiatives for survey, resurvey and settlement;

(b) Modern cadastral mapping, digitization and database creation;

(c) Creation and establishment of National Land Records Modernization Programme Cell, modern records room, State and district level data centres, issues of connectivity and other identified issues and constraints under National Land Records Modernization Programme.

3. To examine issues, challenges and future initiatives:

(a) Reforms initiatives in land records management; legal amendments; and challenges faced in reform process;

(b) Identification of new areas in land administration where ICT application is possible for better governance; and

(c) Development of new policy framework; models and theories; institutional arrangement; and any other innovations in the area of land records management.

Summary of the Full Papers

Chapter 1 on "Current Trends in Land Record Management" by A.A.A. Faizi and H.C. Behera is a discussion about contemporary governance initiatives in the country through institutional reforms by adopting new policies and technology. The chapter discusses various issues in the land records management. This chapter focuses India's position in an international perspective. It provides background about India's e-governance measures in land records management such as strengthening revenue administration and updating of land records (SRA and ULR), and computerization of land records (CLR) and also encompasses discussion about more contemporary reform initiatives through the National Land Records Modernization Programme (NLRMP).

Chapter 2 on "Technology Options for Cadastral Resurvey" by V. Raghu Venkataraman, *et al.* described about various technology options for cadastral survey. They pointed out that a cadastral resurvey in India is pending for a very long period. This would take enormous amount of time, effort and cost with huge manpower. At the threshold of such a massive programme it is very important to choose the right technique, methodology and meticulously formulated standards for smooth progress and acceptance by the people and government. The decision makers and technologists need to have a meaningful discussion on various technology options available to arrive at a consensus. It is very important to make right choices from the various options that are available presently in terms of standards, specifications, procedures, etc. Because these choices would set the direction and pace for the execution of this Herculean task. In the paper the experiences in carrying out the first cadastral resurvey in the country for Nizamabad district in AP through remote sensing techniques are shared. A feasibility study carried out to assess the utility of satellite data for cadastral resurvey has been briefly explained. They pointed out that the suitable methodology for resurvey of a given area should be technically accurate, practically possible and economically feasible. Remote sensing through both satellite and aerial can significantly reduce the cost, time and effort required for cadastral resurvey. CARTOSAT and other High Resolution Satellite (HRS) data have huge potential for prioritization of areas and for updation of cadastral maps. Better accuracy can be achieved with future Indian Missions like CARTOSAT-3 where the spatial resolution is better than 30 cm. For a vast nation with heterogeneous terrain like India, a mixed approach with multiple sensors and techniques would be ideal.

Notes

1. Pankhurst, R.M. (1983): "The Land and The Nation", *LSE Selected Pamphlets*.
2. Bandyopadhyay, R. (1993) Land System in India: A Historical Review, *Economic and Political Weekly*, Vol. 28, No. 52, pp. 149-155.

List of Contributors

A.A.A. Faizi, Professor, CRS, Centre for Rural Studies, LBSNAA, Mussoorie

Charanjit Singh, Director, DoLR, MoRD, Government of India.

D. Sajith Babu (Dr.), Special Officer, Land Bank, Government of Kerala (India).

D.C. Misra, NIC, New Delhi.

D.S. Venkatesh, Principal System Analyst, NIC, DoLR, MoRD.

E. Anantha Padmanabha, Aerial Services and Digital Mapping Area, National Remote Sensing Centre, ISRO.

H.C. Behera, Assistant Professor, Sociological Research Unit, Indian Statistical Institute, Giridih, Jharkhand.

J.S. Ahuja, Advisor, Albireo Telematics Private Limited.

Kishor Mohan Bhattacharyya, Project Director DRISTIDAA kmb191@gmail.com

N.R. Samartharam, Principal System Analyst/Scientist-D,National Informatics Centre, 6th and 7th Floor, Mini towers, Dr. Ambedkar Veedhi, Bangalore-560 00, e-Mail: samarth.nr@nic.in, nrsamarth@hotmail, Phone: 080-22863218, 9449074643, 9480832102

National Informatics Centre

Nivedita P. Haran, IAS, Additional Chief Secretary, Revenue and Disaster Management, Government of Kerala (India).

Prem Singh, IAS, Deputy Director and Centre Director, LBSNAA, Mussoorie.

Rajeev Chawla, Metropolitan Commissioner, Bangalore Metropolitan Regional Development Authority (BMRDA), Bangalore - 560 001,

E-Mail: rajeevchawlaias@gmail.com, rajeevchawlaias@karnataka.gov.in, Phone: 94483 61600

S. Muralikrishnan, Aerial Services mb:and Digital Mapping Area, National Remote Sensing Centre, ISRO.

Srikant Manthri Pragada, Director, Albireo Telematics Private Limited.

Sultan Singh, Senior Scientist, Haryana Space Application Centre (HARSAC), Hisar, Haryana.

Surendra Kumar, IFS, Government of Kerala.

V. Raghu Venkataraman, General Manager, Aerial Services and Digital Mapping Area, National Remote Sensing Centre, ISRO.

Venkata Krishna R., Manager, NIIT-ESRI.

Venkatesan, Deputy Director General and State Informatics Officer, National Informatics Centre, 6th and 7th Floor, Mini towers, Dr. Ambedkar Veedhi, Bangalore - 560 001, e-Mail: sio-kar@nic.in, venkatesan.a@nic.in, Phone: 080-22863790, 080-22863218

Vinay Thakur, Senior Technical Director, DoLR, MoRD, Government of India.

Chapter – 1

Current Trends in Land Records Management

A.A.A. Faizi and *H.C. Behera*

Introduction

India has brought the vision of good governance through various reform initiatives in land records management. Institutional reforms in 1950[1] onwards through land reforms which claimed to have eliminated the process of hegemonic behaviour of the landlords who enjoyed the privilege of power and authority over tenants and cultivators since British colonialism, but the rent seeking behaviour attached with the land records management is still intact. Corruption, litigation, delay in the process of delivery of land related services, lack of transparency, rent seeking behaviour still attached with the manual system (Sinha, 2000). Land services one of the most corrupt sectors in the country reported by Transparency International (2009). There is also lack of clarity in data and the village maps or the cadastral maps found in many States. The maps are old, outdated and not periodically surveyed. Therefore, retrieval of information is difficult. The time of property transaction in South Asia is longest in the world. According to an estimate by World Bank linked site India ranks 94 in terms of efficiency in registering property. The cost involved in the registration process is high which is accounted to be 7.5 per cent of the total property value. The poor governance in land records management has resulted poor return in value of mortgages, poor land investment, land-use pattern and loss of productivity. Ossko

(2011: 27) noted that in developed countries the value of the land and real estate properties, together with mortgages is about 60-65 per cent of the State assets. The land and real estate property related activities generating about 30-35 per cent of the GDP. The value of mortgages on properties in developing countries is generating only 3-8 per cent.[2]

McLaren (2012)[3] says,

> "Only 1.5 billion of the estimated 6 billion land parcels worldwide have land rights formally registered in land administration systems. Many of the 1.1 billion slum dwellers are further billions living under social tenure systems wake up every morning to the threat of eviction. These people are the poor and most vulnerable and are excluded from any form of security of tenure; they are tapped in poverty. Increasing global population and the rush to urbanization is only going to turn this gap into chasm".

Therefore, efforts have been to bring further reforms in the land records management process. First of major reform initiates through governance started way back in 1980s. From the 1980s the system of land records management has been revived by the introduction of information and communication technology (ICT). Computerization is one of the solutions to address problems faced by the Revenue Administration and the landholders. Computerization of Land Records (CLR) and Strengthening Revenue Administration and Updating of Land Records (SRA and ULR) are the emergence of new approaches during 1980s to ensure better management of land records in particular and land administration in general. These are SRA and ULR and CLR. SRA and ULR was meant to modernize survey techniques with the provision of modern equipment such as GIS, GPS, Theodolite, creation of record rooms, purchase of computers and other basic computer accessories, etc., for the speedy survey and settlement activity, updating of land records data and resurvey activities in the regions where surveys were undertaken more than three decades ago. This includes computerization of both spatial and non-spatial data. The spatial data include village map or cadastral map which is in pictorial form and the non-spatial data include textual data like name of the landholders, landholding and account details including crop statistics. Although the objective of computerization of land records is to include both spatial and non-spatial data, yet most States in India have only computerized non-spatial/textual data.

Computerization of land records has been launched across the whole nation in different degree. It has produced mixed result. On positive front, there is increase in efficiency of RoRs and mutation process; quality of RoRs in terms of legibility and clarity are concerned have improved substantially. The process of access and retrieving information both for public and revenue officials have become easier, etc. (Ahuja and Singh, 2003; Chawla, *et al.* 2005; Vachhani and Behera, 2011; Singh and Behera, 2012). The project has got worldwide recognition. Bhoomi project in Karnataka, Tamil Nilam in Tamil Nadu, e-Dhara in Guharat, Bhu-Abhilekh in Madhya Pradesh, Orissa, Haryana; Bhiyan in Chhattisgarh; Bhumi in West Bengal are some of the success stories in LR domain (NIC-LRISD, 2008)[4]. e-Dhara Gujarat is one of the most success stories as far as computerization of textual data and delivery of RoRs are concerned (Ramachandran and Raval, 2008; Vacchani and Behera, 2012). Registration project has also been successfully launched by different States in around 1872 SROs known as CARD, PRISM, Dastavej, ORIS, GARVI, STAR, PEARL, HIMRIS and CORD, etc. (NIC-LRISD, 2008). But the achievements still remain partial as far as the land records as whole is concerned. Digitization of cadastral maps has not been progressed satisfactorily in any of the States except in one or two States like West Bengal. Therefore, the objective of the computerization of land records to integrate both textual and spatial data (maps/graphical outputs) remains unsuccessful in the country.

The major questions asked by the farmers are about location of the lands, the plots, the ownership query, etc. But still there is ambiguity in land related data and in many cases the landholders are not able to be provided with accurate data leading to land related disputes. Therefore, the objectives of land records management has not been largely fulfilled through old schemes such as computerization of land records (CLR) and SRA and ULR. Moreover, the systems of land records are presumptive in nature. Wadhwa (1989:2324) noted that "As regards the Indian law as to title of land, the statute law of India as in force at present, broadly speaking, does not profess to provide for a State certification of title to land under the aegis of a public authority". He quoted the Supreme Court judgment "It is firmly established that the revenue records are not documents of title..." Thus the entries in the land records are challengeable in appropriate court or tribunal (*ibid.*).

The National Land Records Modernization Programme (NLRMP)

In order to address such issues as mentioned above, Government of India introduced the National Land Records Modernisation Programme (NLRMP) in 2008. The NLRMP came into existence in 2008 with the Cabinet's approval for merger of two older schemes namely, the Computerisation of Land Records (CLR) and Strengthening Revenue Administration and Updating of Land Records (SRA and ULR). The Department of Land Resources (DoLR), Government of India defines the aim of the programme is *"to user the process of updating of land records, automated and automatic mutation, integration of both spatial and textual data in land records, interconnectivity of revenue department with the registration department to replace the existing deed registration and presumptive titling system with that of a conclusive titling system with title guarantee in the country"*. The uniqueness in the new programme is to develop a long term vision of establishing conclusive titling system with title insurance, which is called "Torrens System", is being already practiced in Australia, New Zealand, Thailand, Singapore and a few more advanced nations. The principles which is based including single window clearance, mirror principle, curtain principle and title insurance. The concept of single agency to handle land records such as survey and settlement operations, registration of immovable property, mutation, etc., is need of the hour. However, the exercise is predicted as mammoth. It depends on restructuring of the system of land administration. Mirror principle advocates that the land records should reflect ground reality at any given moment. Therefore, the process of resurvey in priority is required. To speed up the activity, intervention of new technology such as application of GIS, SPS, Total Station and aerial survey is encouraged. The curtain principle is again significant where the registration immediately follows mutation without any further enquiry about the past transactions. All that can benefit by lowering transaction cost, hassle free mechanism, etc. Title insurance would ensure government's responsibility to provide title guarantee to the landholders and the party concerned is indemnified against any loss arising because of any inaccuracy in this regard. Provision of guaranteed title to the landholders has been in the forefront among all land related debates by the department of land resources. Ramanathan on 'Land Title Certification' writes:

> The essential aspect to guaranteed title is that it relates to so many aspects of the functioning of both State and markets: it helps in social justice programmes like low income housing; it helps in more effective urban planning, and in the protection of specified land parcels like environmental or heritage assets; it helps faster implementation of infrastructure projects, it reduces delay in judicial processes and unnecessary litigations; it helps in a more efficient mortgage market, with fewer delinquencies and greater transparency; and so on.

With the introduction of NLRMP, the States now have the mandates to effectively execute the NLRMP to facilitate the revenue administration and also to extend benefits to the general public for easy access to land records. However, real benefits will be accrued when creation and updation of land records will be automated and the information will be accurate. Land records should include both the spatial and non-spatial information which should complement each other and maintain high level of accuracy. The Records of Right (RoR) information should reflect the ground reality. The land records should depict both cadastral map and RoR together. Both survey and resurvey have to take place in a specific time frame.

Components of NLRMP

NLRMP has three major components such as computerisation of land records, updation of land records by modern cadastral survey and settlement, and computerisation of registration. The interconnectivity among revenue offices, sub-registrar offices for automatic flow of information, capacity building programmes in addition to a few others are the major additional activities of the programme. NLRMP is certainly important to facilitate the whole process of modern cadastral survey, digitization process, storage and retrieval mechanism, etc. The governance process has to be thoroughly administered only through coordinated effort among various departments. The success of the programme also depends on motivation among all sections of officers and the political representatives concerning land records management.

Activities

- Data entry of textual data, (scanning and digitization),
- Cadastral maps digitization,
- Data centres in tehsils and sub-division level,
- District land records data centres,
- Monitoring cell at State headquarters and training on computer awareness and application software.
- Registration-connection to revenue department, and
- Automatic mutation.

Features of NLRMP

- NLRMP is not only a programme but also an approach towards modernization of land records.
- It can be the basis for introducing conclusive land titling system in the country.
- NLRMP can be useful in bringing commonality in the system of land records management in the long run despites land being a State subject.
- A comprehensive tool for good governance that can ensure transparency, accountability and efficiency in the system of land records management.
- It will facilitate easy access of land data both for citizens as well as planners and management specialists.
- It talks about integration and coordination among the various departments such as survey and settlement, registration, and revenue department who are currently working in isolation.
- A tool for revenue generation through transparency property transaction, accurate property location, and valuation of property, etc.
- An approach for sustainable development.

Current Issues and Challenges

Land is a critical component for survival, growth and development. Therefore, irrespective of nature and position of land, its management is crucial. However, effective management is possible only when the land positioning, size, ownership details, quality and nature of soil, etc., are properly surveyed. India is a typical case of land management based on its cultural pluralism, co-existence of both local and global land record management practice. Customary land tenure system with multiple uncodified laws and principles exists in some parts of the country particularly among indigenous communities.

In many areas, fresh cadastral survey has to be started which seems to be a daunting task. There has to be creation of revenue codes and guidelines with attention to local sensitiveness, local culture and practices. On the other hand, the change in land market relation, increase in pressure on land revenue administration due to development agenda, etc., have led conflict. Much change is noticed in recent years due to change in land-use pattern. As per the change noticed, there is need for change in the old maps for the areas where there is substantial change noticed, etc. These technical problems may seriously damage the credibility of land reform measures. The application of ICT and other modern technologies need to be extended with adequate institutional mechanism. Currently, many States, majority from the north-eastern region, have to take this mammoth exercise afresh. That has to be coherent with traditional customary practices as well. Intensive study is required in the State where role of community, civil society organisation, and local administration is important. The detailed project report has to be prepared after understanding local situation and demand for modern technology application. The State Government alone cannot take up this initiative. The public and private partnership required with a PPP model. For example for application of GIS, Remote Sensing and other modern technologies for survey, resurvey and cadastral mapping, large number of private agencies and corporate bodies in India are well equipped. They have adequate manpower, technology, models and approaches and vision. However, their roles have to be identified. Similarly, digitisation process requires outsourcing from private parties. This requires a baseline study. Since, land is a legal subject and the revenue administration is more complex in nature, the disputes may easily take place. This may cause delay in operation and push the process of reform further in the backtrack. Therefore, the role of scientists and social scientists will be equally important for the entire operational mechanisms. The PPP models have to be examined. Various components of the PPP model such as lease, transfer, service contract, concession agreement, divesture, etc., have to be studied thoroughly. The World Bank initiative to study the process and mechanism of PPP model in land records management can help further study.

Implementation Challenges: Field Observations

Land Records Management system is immensely diversified in all the states of India. The type of land records and the number of registers vary across States. For example, *jamabandi* in Uttar Pradesh, 7/12, 8A and VF-6 in Maharashtra and Gujarat; *khatauni* in Odisha, *chitta* in Tamil Nadu, etc. Similarly, the types of maps such as *tippan*, *hal*, *mauza* maps, etc., vary across States. Some States have map sheets for plots (*Tippan*). For example, in Maharashtra there are *Tippans*. Due to plot maps, the number of maps in the State is large in comparison to the States where there is only village/mauza maps which are large in numbers. West Bengal and Odisha have village maps/*mauza* maps, *hal* maps, etc.

The scanning and digitization of old land records is a daunting task. The cost for scanning varies depends on the number and size of maps and also the condition of maps. Digitization of maps after seamless mosaicing often leads to mismatch in the area and size of plot maps with that of the original plot maps.

- Already existing huge backlog data for mutation thus resulting delay in updation.
- Litigation and land related disputes.
- Shortage of manpower including surveyors.

Currently, many States, majority from the north-eastern region, have to start afresh this mammoth exercise, one which is coherent with traditional customary practices as well. Intensive study is required in a State where the role of the community, civil society organization and local administration is important. The detailed project report has to be prepared after understanding local situation and demand for modern technology application. The State Government alone cannot take up this initiative, what is required is public and private partnership, PPP model. For example, for application of GIS, Remote Sensing and modern technologies for survey, resurvey and cadastral mapping, large number of private agencies and corporate bodies are well equipped in our country. They have adequate manpower, technology, models and vision. However, their roles have to be identified. Similarly, digitization process requires outsourcing from private parties which requires a baseline study. Since land is a legal subject with revenue administration being complex, disputes may arise which lead to delay in operation and slow the entire process down. Therefore, the role of scientists and social scientists will be equally important in the operational mechanism. The PPP models have to be examined with its various components such as lease, transfer, service contract, concession agreement, divesture, etc., to be studied thoroughly. The World Bank initiative to study the process and mechanism of PPP model in land records management can help further study.

Development Vision

Despite many challenges before true implementation of the NLRMP, it can influence land market and induced development intervention if properly executed. Financial and credit institutions can be active. There will be better disaster management and rehabilitation and resettlement operation, etc., the impact can be seen in the cropping pattern, land market relation, industrial production, etc., the landholders will be benefited both directly and indirectly with real time land records in hand and State can prosper.

The introduction of Bills such as land titling, land acquisition, rehabilitation and resettlement (LARR) Bill are not isolated initiatives but can be effective only with emphasis on the National Land Records Modernisation programme. However, both law and administration should move step by step to address various uncertainties and issues pertaining to land records management. Public awareness should be conjoined in the line of development intervention. Capacity building measures at various departments through training, field visit/exposure visit, workshops and seminars need to be done with equal proportion.

Notes

1. The land reform policy was concretized for the first time in the First Five Year Plan. Agrarian classes were classified into intermediaries, large owners, small and medium owners, tenants at will and landless workers. Special emphasis was given on the abolition of intermediaries, enactment of radical tenancy laws and resumption of cultivation based peasant proprietorship (Bandyopadhya, R. (1993).
2. Ossko, Andras (2011) Need for New Services in Land Administration, Coordinates, Vol. VII, Issue 7.
3. Crowd sourcing and land administration, Coordinates. Retrieved from http://mycoordinates.org/crowdsourcing-and-land-administration/.
4. Technical Report on Study of Uniform Coding Scheme for Computerisation of Land Records, NIC-LRISD-001. Retrieved on 9th October 2012 from http://www.dolr.nic.in/dolr/downloads/pdfs/lrc_codes_list_finaldraft300908reviseda.pdf.

Chapter – 2

Technology Options for Cadastral Resurvey

V. Raghu Venkataraman, E. Anantha Padmanabha
and *S. Muralikrishnan*

Introduction

In India the main purpose of cadastral system was to generate revenue through property tax collection. (Prithvish Nag, 2003). There is a paradigm shift with increasing value of land, the legal ownership has become important now. Depiction of legal and topographical aspects of the land becomes more important if it pertains to urban/town property. For developmental activities to go at a fast pace, the cadastral information is essential for administrators, policy and decision makers. Lack of such information becomes a bottleneck for any development activity and can lead to legal and political issues. Inaccessible cadastral information also leads to insecurity among the people, which becomes an advantage for anti-social elements. There is a consensus across the sections of people that a cadastral information system can add momentum to the nation's progress in development and can reduce the burden on our legal system.

A cadastral resurvey in India is pending for a very long period. This would take enormous amount of time, effort and cost with huge manpower. There are multiple methods by which a Cadastral resurvey can be taken up. Traditional ground based methods include plane table survey using chain, tape or theodolite, Total station survey, and GPS surveys. Remote Sensing techniques like Photogrametry from aerial and satellite platforms can also be used for cadastral resurvey.

Earlier cadastral survey in India was done using ground survey techniques. For small areas the ground based surveys can be used for accurate and economical cadastral survey. But for cadastral survey of large areas such as towns, districts and States traditional method is not feasible. As the ground survey of large areas require huge manpower working in the field, irrespective of the weather conditions for months and years. With more personnel involved over a long span of time chances of committing errors would be more. These limitations have been minimized with the introduction of advanced survey equipment like Electronic Distance Measurement (EDMs), Total Stations and Global Navigation Satellite System (GNSS), etc. But they have not been fully eliminated.

Now with State of the art technologies like Remote Sensing from aerial and satellite platforms, the time, effort and cost required would be significantly reduced, with minimum manpower. In the last couple of decades the Remote Sensing has undergone numerous advancements in terms of data acquisition and processing. Through Remote Sensing larger areas can be surveyed in shorter span of time. Most of the cadastral survey can be carried out with comforts in the laboratory, with fieldwork minimized to large extent. Unlike the conventional ground based methods the outputs from Remote Sensing can be directly used in a GIS based land information system.

In spite of significant technical advancements, the cadastral resurvey for India is going to be a Herculean task with many technical challenges. Total area of the nation is 3.29 million sq. km and the area to be surveyed is approximately 2.16 million sq. km with around 600 districts in 28 States and 7 union territories. The Cadastral resurvey of such a large area could probably be the largest exercise in the world (Rita Sinha, 2008). At the threshold of such a massive programme it is very important to choose the right technique, methodology and meticulously formulated standards for smooth progress and acceptance by the people and government. Moving in the right direction at this point is very critical otherwise that may lead to a never ending loop.

Survey Techniques: Making the Right Choice

The choice of suitable technique for a fresh cadastral resurvey has led to the debate among the decision makers and the technical community. Ideally the survey technique that is to be adopted should be technically accurate, practically possible and economically feasible. The possible techniques by which a fresh cadastral resurvey can be taken up are (i) Digitization of existing records, (ii) Ground based technique using Totalisation and GNSS or (iii) Remote Sensing from aerial or satellite platforms (Fig. 2.1). Each of the above mentioned techniques have advantages and limitations. A comparison of these techniques is given in Table 2.1.

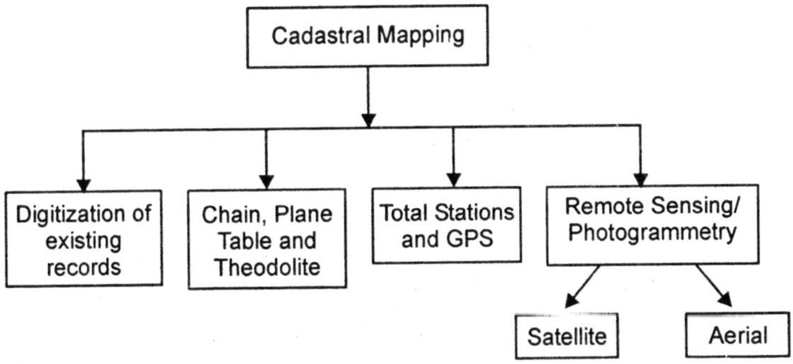

Fig. 2.1: Cadastral Mapping Methods

Before taking up this massive exercise, the decision makers and technologists need to have a meaningful discussion on various technology options available to arrive at a consensus. It is very important to make right choices from the various options that are available presently in terms of standards, specifications, procedures, etc. Because these choices would set the direction and pace for the execution of this Herculean task.

The process of selecting the best suitable technique depends on:

- Existing records and their condition.
- Budget and Time.
- Area extent to be surveyed.
- Capacity available.
- Accuracy required.
- Terrain conditions.
- Land value.
- Land-use (built up, forest, waterbody, etc.).

Table 2.1: Comparison of Cadastral Mapping Methods

Item	*Chain, Plane Table, Theodolite*	*Total Station and GPS*	*Satellite*	*Aerial*
Datum	Local	Local/WGS 84	WGS 84	WGS 84
Field survey for vertices	Required	Required	Required	Required
Measurements and plotting	Field survey for length and height	Field survey for length and height	Laboratory	Laboratory
Controls	Arbitrary control	Link to bench marks	DGPS	DGPS
Process	Plotting	Plotting	Triangulation, BA, DEM extraction, ortho generation, tiling	Triangulation, BA, DEM extraction, ortho generation, tiling
Validation	Field visit	Field visit	Laboratory— Since image evidence is available	Laboratory— Since image evidence is available
Post-processing and adding attribute data	Inbuilt	Inbuilt	From field data	From field data
Accuracy subject to point identification	1 link per chain = 30 cms.	5 mm + 1 ppm	Likely 1 pixel	Upto 10 cm
Suitability	Small areas	Slightly large areas	Very large areas	Large areas
Manpower	Very huge	huge	Very less	less

For the areas where the survey records are in good condition, digitizing the existing records can be the most economical and simple solution. For this the primary requirement is that the existing records should be intact and digitizable. These records after digitization can be georeferenced to rectified images from high resolution satellite sensors. Only the problematic areas can be resurveyed either with total station or GNSS or both. But the datum and projection details of the existing records may not have been well defined. So there could be problems in georeferencing the old records and may lead to distortion if forced to overlay on to the satellite data set. There could be edge matching problems with the areas that require resurvey.

For smaller areas Total station can be effectively and accurately used to demarcate the property boundaries. For large area surveys more instruments and more personnel are required as each and every property vertex is to be visited manually. There is also a possibility of committing errors as more personnel are involved in the work for longer durations. A clear line of sight is required and hence operation is not possible in poor light conditions. A GNSS on the other hand doesn't require clear line of sight and can be operated in poor light conditions and during night. But a clear sky away from any obstructions and tall structures is required for GNSS survey as the signals are to be received from satellites. As each vertex of the property boundary should be visited manually, signals may not be received at many places due to obstructions.

Remote Sensing images for the cadastral survey may be acquired from both aerial and satellite platforms. These images have to be processed through photogrammetric procedures to generate the cadastral maps. Remote Sensing technique can save time, cost and effort, as there is no need to visit each and every property vertex manually. Property boundaries can be demarcated in the laboratory with minimum field visits and with less manpower.

Through images acquired from the aerial platform, an accuracy of upto 10 cm (with images at 5 cm resolution) in distance measurement can be obtained. But the aerial data acquisition is very slow and is dependent on weather and clear visibility. Also the capacity and infrastructure in the country is limited and not sufficient at the moment and needs to be augmented. Data acquisition from satellite is faster compared to aircraft but is dependent on clear visibility.

Presently the best possible spatial resolution from satellite is 50 cm and hopefully in the near future the spatial resolution may be further improved. For rural areas where the land value is very less compared to the urban areas, the existing satellite data can be used for cadastral resurvey provided the accuracy goals are set realistically. Satellite data can also be used to demarcate the forests, waterbodies and inaccessible terrains. But for urban areas the land value is very high and the cadastral survey needs to be very accurate, otherwise that may lead to serious property disputes and resistance from the people. But for satellite data, to be used for urban cadastre, there is still a gap of few centimeters, which needs to be filled up.

For the best and feasible cadastral resurvey methodology, satellite data may be used for rural areas with large land parcels and remote areas. For urban areas aerial data should be used to meet the accuracy requirement. The minimum control required for both aerial and satellite data can be obtained from GNSS. All the property vertices cannot be mapped from either aerial or satellite data. Places where the property vertices are not visible in the Remote Sensing data due to obstructions or relief, Total station survey may be carried out to fill up the gap areas. Thus it can be concluded that no single technique can give complete solution for any vast country with heterogeneous characteristics. A multi sensor and multi technique approach should be adopted for effective results, which would become a landmark in the survey history.

Recent Trends in Survey Technology

The Remote Sensing technique serves a number of advantages compared to conventional methods of mapping the topographic surface of the earth. Comprehensive advantages include significant saving in cost, time, and efforts that needs to be put in preparing maps. Topographic data can be used as an effective input for decision making in various spatial applications. Cadastral application can be a classic example that can exploit the maximum potential of the Remote Sensing both from aerial and satellite platforms.

Remote Sensing in the last couple of decades has undergone numerous changes and is evolving as an undisputed replacement for the conventional ground based surveying method. These changes in the form of advancements made an obvious difference in every phase of mapping. Starting with data acquisition from aerial to satellite platforms at rapidly improving resolutions, i.e., spatial, spectral, radiometric and temporal. The transition from the analogue film based cameras to the digital cameras of large format for covering large areas. The use of multiple sensors onboard like GNSS, IMU along with Large Format Digital Cameras, for ready to use Exterior orientation parameters. Without GNSS for control survey the accuracy in cm for large areas was unimaginable. High end storage and processing systems with fully automated software have reduced the processing time drastically. These systems are capable of handling huge volumes of data and can generate customized products in desired format with ease.

On the satellite platform side during last decade quite a good number of satellites have been launched with a spatial resolution of 1 m and better. A brief list of details is given in Table 2.2. Hopefully future missions may reduce the gap in spatial resolution between aerial and satellite data. Especially CARTOSAT-3 with a proposed spatial resolution better than 30 cm, to be launched by ISRO, primarily designed to meet the cadastral resurvey demand. With satellite data the effort and time required for data acquisition and processing is significantly less compared to that of aerial. Satellite data because of higher altitude of acquisition will be less affected by the relief displacement than aerial platform. For rural areas where the land value is very less compared to the urban areas, the existing satellite data can be very well used for cadastral resurvey provided the accuracy goals are set realistically. Satellite data can also be used to demarcate the forests, waterbodies and inaccessible terrains.

The minimum ground control required can be obtained through differential GNSS survey. The invention of GNSS has simplified entire ground control survey process, which used to be the most tedious part of any mapping project. GNSS can be used to observe the control network with points separated by kilometres of distance, as the survey does not require the line of sight. GNSS survey can be carried out in any weather condition and both during day and night thus reducing the time spent in the field.

The improvements in the hardware and software industry is at a very fast pace. Today it is possible to store and process the amount of

Table 2.2: High Resolution Satellite Sensors

	Ikonos	*Quickbird*	*Cartosat-2/2B*	*Worldview-2*	*Worldview-1*	*GeoEye-1*
Launch	Sep. 2, 1999	Oct. 18, 2001	Jan. 12, 2007 July 12, 2010	Oct. 8, 2009	Sept. 18, 2007	Sept. 6, 2008
Pixel size	PAN – 1 m MX – 4 m	PAN–0.61 m MX– 2.44 m	PAN – 0.8 m	PAN – 0.46 m MX – 1.84 m	PAN – 0.5 m	PAN-0.41 m MX-1.65 m
Swath	11 km	16.8 km	9.6 km	16.4 km	17.6 km	15.2 km
Quantization	11 bit	11 bit	10 bit	11 bit	11 bit	11 bit
Stereo Acquisition	Yes	Yes	Yes	Yes	Yes	Yes

data that was considered impossible a couple of years ago. Systems capable of handling high resolution image data covering cities, districts, States and countries are available for the geospatial community. Because of such systems geospatial wonders like Google earth and Bhuvan have become reality today and are available at the fingertips of the people.

All these technological advancements can be effectively exploited to reduce the time, cost and effort required for the cadastral resurvey of the entire nation. Things have changed a lot compared to the time when the last cadastral survey was done. It is beyond doubt that the industry and geospatial community is better equipped presently to carry out this massive task.

Nizamabad Experience

The Bhubharati project as a pilot study was initiated by Government of Andhra Pradesh for cadastral mapping of Nizamabad district. For the first time in the country an entire district was taken up for cadastral resurvey through Remote Sensing. The aerial survey and photogrammetric processing was taken up by NRSC, ISRO, Hyderabad. The technical details of this project are given in Table 2.3.

Table 2.3: Technical Details of Bhubharati Project

Project	*Bhubharati*
Location	Rural–Nizamabad district, Urban–six towns in Nizamabad area.
Area of acquisition	Rural– 10,000 sq km, Urban – 390 sq km.
Area of mapping	Rural–6300 sq km, Urban – 390 sq km.
Photography scale	Rural–1: 10000, Urban – 1: 4000
Product	Orthoimage (Rural– 15 cm, urban – 5 cm GSD)
Accuracy	Rural – 25 cm, Urban 10 cm, in distance measurement.
User	Government of Andhra Pradesh.

Aerial data was acquired with a large format file based camera RMK top, at 1 : 10000 scale for entire district and at 1 : 4000 scale for six urban areas. The district boundary along with the location of six urban towns is shown in Fig. 2.2. The exposed film was chemically processed and then scanned to convert the photographs into digital images. Ground control survey was carried out through differential GNSS for the required control in block adjustment. Digital images,

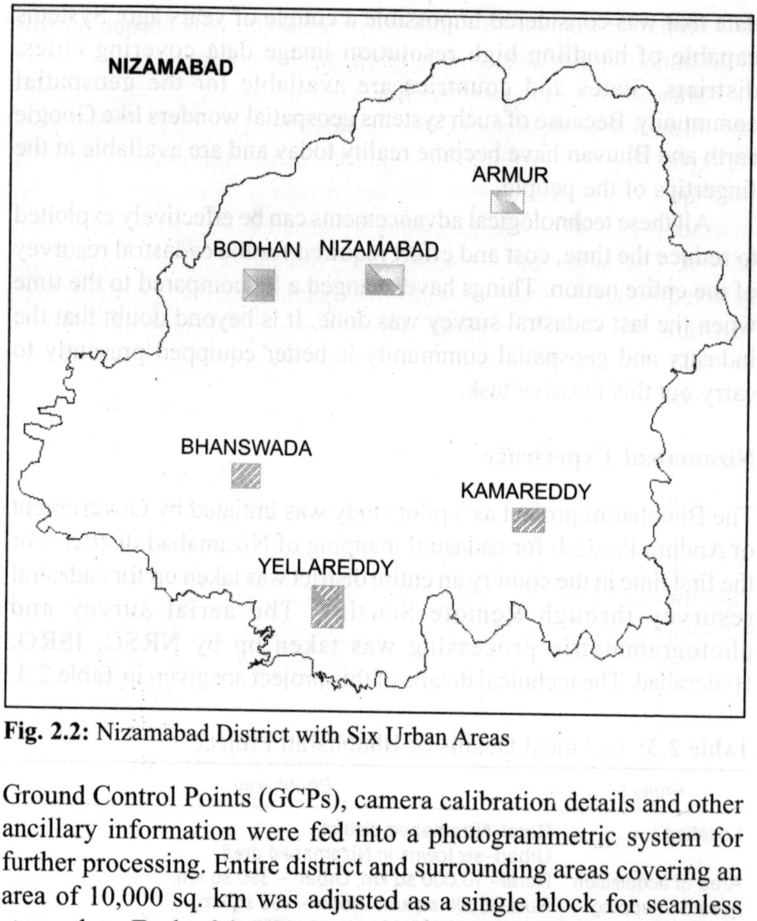

Fig. 2.2: Nizamabad District with Six Urban Areas

Ground Control Points (GCPs), camera calibration details and other ancillary information were fed into a photogrammetric system for further processing. Entire district and surrounding areas covering an area of 10,000 sq. km was adjusted as a single block for seamless stereo data. Each of the six urban areas was also adjusted separately. Digital Terrain Models (DTM) was generated from the adjusted stereo data. These DTMs were used for orthorectification of all the adjusted images. Individual orthoimages were mosaicked for uniform and seamless appearance and orthotiles of size 1000 m × 500 m were generated.

Before acceptance, the final product, i.e., orthoimage tiles were validated in the field with reference to measurements made using Total station. Utmost care is taken in selection of points so that the variation in repeated measurements and by different personnel is minimum. Points that are geometrically sharp both on the ground and on the orthoimage like culvert corner, water tank corner, narrow field bund junction, etc., are selected for the study.

Sharp points were selected on the orthoimage at different villages spread across the district for distance measurement. These points were identified on the ground and marked on the hardcopy maps for subsequent measurement in the laboratory using orthoimage. Horizontal distance between the selected points is measured using Electronic Total station. The horizontal distances observed in the field between the selected points are measured on the digital orthoimage using CAD software. The difference between the horizontal distances observed in the field and the horizontal distances measured on the digital orthoimage in the laboratory were found to be within the limits, i.e., 25 cm in rural areas and 10 cm in urban areas with 90 per cent confidence.

Study on Feasibility of Satellite Data for Cadastral Mapping

A feasibility study was carried out to assess the utility of satellite data for cadastral resurvey. Following high resolution satellite data sets were used for the study.

1. Cartosat-1 stereo data of 2.5 m resolution,
2. Cartosat-2 mono data of 0.8 m resolution,
3. Ikonos mono (True Colour) 1 m resolution,
4. Quickbird mono (True Colour) 0.6 m resolution,
5. Aerial orthophoto of 15 cm resolution, and
6. Aerial DEM 1 m accuracy.

For this study, the aerial orthoimages and high resolution satellite data of CARTOSAT-1, CARTOSAT-2, IKONOS and Quickbird were used. Basically the aerial ortho and stereo models generated earlier are kept as reference for comparison with satellite data. The CARTOSAT-1 ORTHOKIT stereo data is processed using Rational Polynomial Coefficient (RPC) model provided with the data. The RPC model provides the relationship between object space coordinates (latitude, longitude, height) to image space (line, sample) coordinates. The triangulation of the CARTOSAT-1 data is processed using RPC model, and the refinement of the model is carried with GCPs. After triangulation, Mass points were collected with 50 m spacing and break lines were collected wherever there is change in terrain as well as along streams, ridges, etc. Ortho image is generated from aft data using the DTM derived from CARTOSAT-1 stereo data set. (Narender *et al.*) Similarly other satellite data sets were georeferenced using GCPs that are uniformly distributed over the study area. DTM generated from CARTOSAT-1 data was used for orthorectification of the other satellite data sets. Distance and area measurements were made using the orthoimages generated from satellite data sets and compared with aerial orthoimages. An accuracy of around 1 pixel in distance measurement and 2 per cent in area measurements were achieved.

Conclusions

1. The suitable methodology for resurvey of a given area should be technically accurate, practically possible and economically feasible.
2. If existing records are intact, digitizing them would be the most economical and simple solution.
3. Cadastral resurvey through ground surveys require lot of time, money and manpower.
4. Remote Sensing through both satellite and aerial can significantly reduce the cost, time and effort required for cadastral resurvey.
5. CARTOSAT and other HR data have huge potential for prioritization of areas and for updation of cadastral maps.
6. Better accuracy can be achieved with future Indian Missions like CARTOSAT-3 where the GSD is better than 30 cm.
7. For a vast nation with heterogeneous terrain like India, a mixed approach with multiple sensors and techniques would be ideal.

References

Abdullah M. Al-garni, 1996; "Urban photogrammetric data base for multi-purpose cadastral-based information systems: the Riyadh city case", ISPRS *Journal of Photogrammetry and Remote Sensing*, 51, pp. 28-38.

Behl K.G., 2003; "Cadastral map specification, methodology and infrastructure for adoption by the States", *Indian Cartographer*.

Cheng Nga-fong, Tang Hong-wai Conrad, 2002, "An Analysis and implementation of cadastral survey system in Hongkong", symposium on geospatial theory, processing and applications, Ottawa.

Eugence H. Silayo, 2005; "Searching for an affordable and acceptable cadastral survey method", From Pharaohs to Geoinformatics–FIG working week, Cairo, Egypt.

Igor Jaramillo, 2000; "Mathematical rectification of aerial photographs for cadastral mapping in Bolivia", *International Archives of Photogrammetry and Remote Sensing*, Vol. XXXIII, part B4, Amsterdam.

Liang-Hwei Lee and Tsu-Tse Su, 1996, "Vision-based image processing of digitized cadastral maps", *Photogrammetric Engineering and Remote Sensing*, Vol. 62, No. 5, pp. 533-538.

Narender B. Jayalakshmi I., Nalini J. and Ananatha Padmanabha E., S. Muralikrishnan and V. Raghuvenkataraman, 2008; "Geometric rectification of high resolution space data". Workshop on Computer and Information Technology at VSSC, Trivandrum, India.

Paul S. Anderson, 2000; "Mapping land rights in Mozambique", *Photogrammetric Engineering and Remote Sensing*, Vol. 66, No. 6, pp. 769-775.

Prithvish Nag, 2003; "Cadastral Country Report 2003".

Prudhvi Raju K.N., Sarvajeet Kumar, Kshitij Mohan and Manish Kumar Pandey, 2009; "Urban cadstral mapping using very high resolution Remote Sensing data", *Journal of the Indian Society of Remote Sensing*, Vol. 36, No. 3, pp. 283-288.

Raghu Venkataraman V., Jayalakshmi I., Srinivas P., 2008; "Digital ortho photo generation using large scale aerial photography for cadastral resurvey of Nizamabad district in Andhra Pradesh, India". International Symposium, on Contemporary Trends in Cadastral Surveys and Land Titling, Hyderabad, India.

Rao D.P., Navalgund R.R., Krishna Murthy Y.V.N., 1996; "Cadastral applications using IRS 1C Data— Some case studies", *Current Science*, Vol. 70, No. 7.

Rita Sinha, 2008; "Moving towards clear land titles in India: Potential benefits, a road-map and Remaining challenges". Innovations in land rights recognition, administration, and governance. available at: www.issuu.com/worldbank.publications/docs/9780821385807.

Chapter – 3

Aerial Survey and Computerized Web Based Land Information System

Srikanth Manthripragada and *J.S. Ahuja*

Background

Land, as a resource, belongs to some individual or group of individuals having legal rights to own the land. These individuals are free to use their land according to their will or to sale their land for monetary or other advantages. Therefore, access to land is an important factor to reduce poverty and create economic growth. 'But the problem is that though the land belongs to an individual but land related issues like sale-purchase registration, granting a mortgage, to impose or lift restrictions and to allocate a certain land-use comes in the domain of the Government. The administration having all the discretionary powers and loopholes in the laws to grant the formal sanctions in all these land related issues considers them as personal favours to the landholders giving rise to mass scale corruption. This phenomenon of land related corruption is very common throughout the world. It is not restricted to third world countries.

Encroachments

The updating of land records has been the most neglected area of our governing system. Even the Government, the biggest landholder, does not know their coordinates. Encroachment, particularly the forest land

and the rural areas on the periphery of urban agglomeration, has been on the increase in the free India, breeding corruption. This act is mainly done with the blessings of people to reward their associates supporting them to come to power. The chart shows the forest areas under encroachment. The main reason for committing this type of offences has been non-availability of updated land records. The State-wise statistics of encroachment of forest land published very recently in the national press is reproduced.

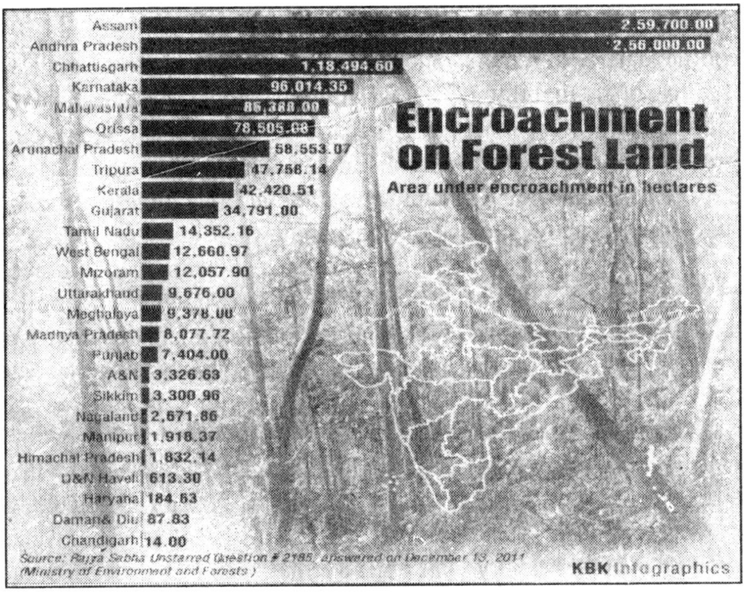

Fig. 3.1: Encroachment on Forest Land

This has happened though India with an area of 6,35,400 km^2, constituting about 19.32 per cent of the land area of the country, constituted the Imperial Forest Department as early as in 1864. But no separate cadre of Land Managers was created by the British Raj that time and there is no change in the situation.

National Land Records Management System

Majority of the developmental activities has always been land based for which identification and acquisition of the land parcels is the starting point. Identification means its location, coordinates, owners and other details. All these informations form part of Cadastral Records/Revenue Surveys. Land has, therefore, been one of the most important and controversial subject in the country when it comes to both its identification and its acquisition. This task can be best done if there is a proper record of land maintained in descriptive and spatial form by the State Governments. But the land survey/resurvey system has been under neglect and outdated in most of the States of the country, adversely affecting the GDP.

Registration of land though has become a good source of revenue for the State Governments but its full potential are not being realized due to non-availability of spatial and numerical data and thematic information. On the other hand, the land owners too have been facing problems during land acquisition process due to lack of coordination between the multiple agencies, presumptive titles, time consuming mutation and above all lack of transparency.

Now when we look at the current status of techniques employed to survey/resurvey of land by the Revenue Departments, it does offer a lot of scope for change in the total system. Realizing the gravity of the situation Government of India had directed the States to take up Survey/Resurvey of lands for under their National Land Records Modernization Programme (NLRMP) to modernize management of land records, minimize scope of land/property disputes, enhance transparency in the land records maintenance system, and facilitate moving eventually towards guaranteed conclusive titles to immovable properties in the country. This objective has been well brought out in their document issued by Department of Land Resources, Ministry of Rural Development, Government of India. This document contains various case studies based on the experiments carried out in different States. A proper NLRMS will provide the following benefits to the land owners, a right granted by the Constitution to the land owners:

- Guarantee of ownership and security of tenure;
- Reduce land disputes;
- Make the Land Acquisition task easy for urban planning and infrastructure development;
- Support for land and properly taxation;
- Provide security for credit;

- Facilitate rural land reforms;
- Support environmental management; and
- Produce statistical data.

With this background, Department of Land Resources, Ministry of Rural Development, Government of India, has launched National Land Records Modernization Programme (NLRMP) for updating of land records in the country. This programme envisages generation of a complete cadastre for the entire country for all private and State land parcels. This cadastre has to be geo-referenced in terms of the geodetic datum based on WGS 84. The cadastre generated will have to be spatially consistent meeting the legal requirements. The ultimate goal of the NLRMP has been to have a conclusive titling system with title guarantee, to replace the current presumptive title system in the country. Every land owner private as well as Government expects the system to grant them with the title of the land and to reduce land disputes as part of good governance.

Scope of the Programme

The major component of this NLRMP has been to survey/resurvey of the urban, village and agriculture areas to prepare digital maps by survey/resurvey on different scales and accuracies.

Table 3.1: Scale and Level of Accuracy in Different areas under NLRMP

Area	*Scale*	*Accuracy to be Achieved/Specified in the State Land Survey Act*
Urban Areas	1 : 500 to 1 : 1000	± 5 cm
Village Areas with population	1 : 1,000 to 1 : 4000	± 10 cm
Agricultural Land	1 : 4,000 to 1 : 8000	± 20 cm

The survey/resurvey in this programme has to be carried out in absolute terms based on national grid with the aim to prepare Land Parcel Maps in GIS environment with coordinates – Latitude, Longitude and Altitude to give them uniqueness. This long drawn and a very ambitious scope of the programme has been as under. Their ultimate goal is to make feasible the conclusive titling system with title guarantee, to replace the current presumptive title system in the country.

Computerization of Land Records

(a) Data entry/re-entry/data conversion of all textual records including mutation records and other land attributes data;
(b) Digitization of cadastral maps;
(c) Integration of textual and spatial data;
(d) Tehsil, sub-division/district Computer centres; and
(e) State-level data centres and Inter-connectivity among revenue offices.

Survey/Resurvey and Updating of the Land Records

There are basically two accepted methods to prepare such large scale maps with required accuracy, being legal documents. One is by ground survey and the other by using ground image such as Aerial Photography/Satellite Imagery. But the technology of Aerial Photogrammetry by flying the area for photography supplemented by ground control with Electronic Total Stations (ETS) and Differential Global Positioning System (DGPS) may give us Scale, accuracy, speed and cost.

Both the two broad methods include establishment of network of ground control points (GCP) and ground truthing and have further divided into three options:

(a) Pure ground method using total station (TS) and differential global positioning system (DGPS);
(b) Hybrid methodology using Aerial Photography and ground truthing by TS and DGPS; and
(c) High Resolution Satellite Imagery (HRSI) and ground truthing by TS and DGPS.

Establishment of network of Primary, Secondary and Tertiary Ground Control Points (GCPs) and their monumentation is common to all the three methodologies.

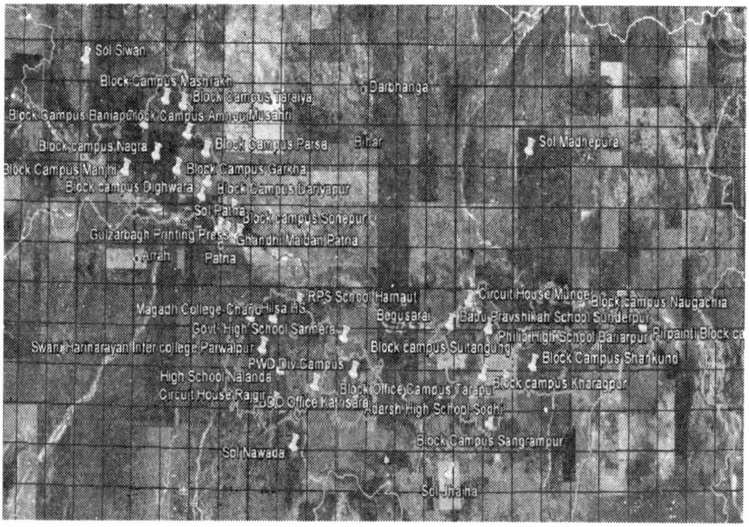

Fig. 3.2: Network of Ground Control Points

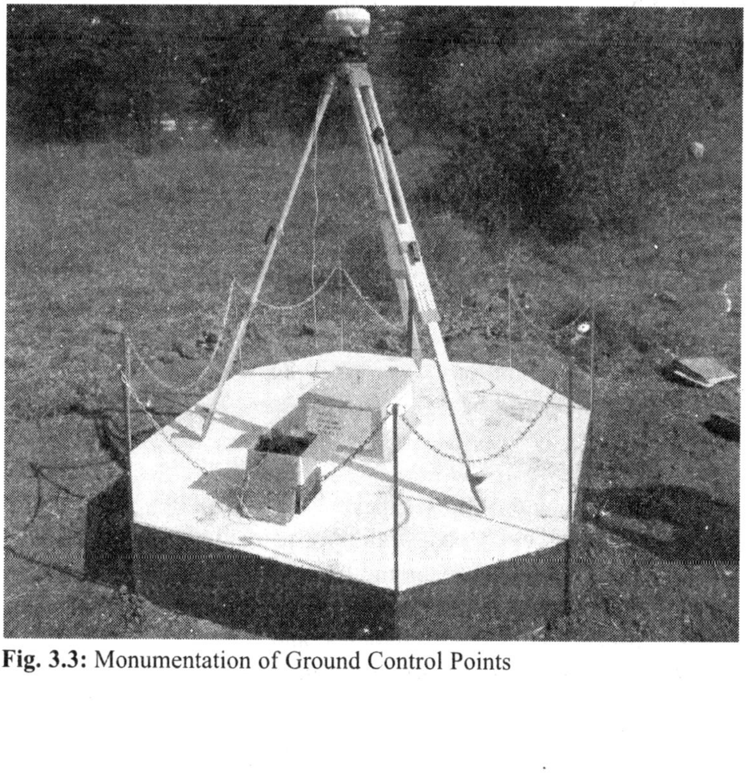

Fig. 3.3: Monumentation of Ground Control Points

The comparison between these three technologies taking into considerations all the factors including the basic four factors of Scale, Accuracy, Speed and Cost has been given in Appendix l.

This will enable the States to decide on the technology to be chosen for this project of national importance.

Computerization of Registration

(a) Computerization of the Sub-Registrar's Offices (SROs);
(b) Data entry of valuation details;
(c) Data entry of legacy encumbrance data;
(d) Scanning and preservation of old documents; and
(e) Connectivity of SROs with revenue offices.

The Technical Manuals were prepared after obtaining inputs from the leading technical agencies as well as from the field experience of States which have done commendable work in specific areas of the programme. Detailed Guidelines and Technical Manual have been issued to all the Land Survey and Settlement Departments of State Governments for better implementation of the NLRMP. This has now been left to the State Governments to initiate action to implement this centrally sponsored programme.

Requirements

A complete cadastre is, therefore, required to be generated for the entire country partly through a centrally funded project, for all private and State land parcels. This cadastre has to be geo-referenced in terms of the geodetic datum based on WGS 84. The cadastre generated under this project will be:

- Spatially consistent;
- Current and comprehensive;
- Each and every cadastral lot or parcel uniquely identified;
- Relationship of each land parcel with any other abutting or adjoining land parcel;
- Meets the legal requirements; and
- Accuracies prescribed are achieved.

Preparation of Modern Cadastres

Cadastral Surveys establishes the most important link of ownership between man and the land giving them the record of rights through a document known by different names in the country as khasra, khatauni, mouza and others. It involves the depiction of legal and topographical aspects including the precise area of their land parcels. The outdated and in many cases inaccurate surveys and cadastral records are the major bottlenecks in carrying out land reforms in rural areas for their overall development. It also makes the task of acquisition of land difficult and time consuming, when required by Planners and Engineers for activities even at national level.

Comparative Analysis of Three Technologies

Ground method is cumbersome, labour intensive, time consuming, denser GCP requirement dictated by ground conditions like and require extensive field visits by Surveyors for ground truthing. It may take years to clear the backlog. The very high resolution data from IKONOS, CARTOSAT and QUICKBIRD available with resolutions of 1 metre, 2.5 metres and 0.60 metre respectively, cannot generate the cadastral maps of the required accuracy considering the nature of landholdings. Technologies like High resolution Aerial stereo imagery with LiDAR data has potential to replace traditional Ground Survey techniques. These technologies has gained a major importance in creation of DEM and Base Maps in the western parts of the World

Comparison between Satellite Imagery and Aerial Photo

The accuracy achieved for Geoeye test data, shows that the Map Scale of 1 : 7000 and above is achievable as per ASPRS standards reference (Table) for Aerial photo, but the problem lies with the GSD of the satellite Data, for Geoeye has a GSD 0.5 m

A target scale of 1 : 6000 and less is not possible with resolution 0.5 m as some of the finer features like poles, manholes will not be visible on these satellite images and, therefore, for such map scales Aerial photo is used. With the present commercial stereo satellite imagery available a target scale of 1 : 7000 or more is possible.

Table 3.2: Difference of Aerial Photography and Satellite Images on difference Ground Sample Distance or GSD

	Target Scale	*Aerial*	*Satellite*
AT GSD 10 CM	1:500	0.05 m	Not Good
	1:1,000	0.05 m	Not Good
AT GSD 20 CM	1:2,000	0.10 m	Not Good
	1:2,500	0.10 m	Not Good
	1:3,000	0.10 m	Not Good
	1:4,000	0.10 m	Not Good
	1:5,000	0.10 m	Not Good
AT GSD 50 CM	1:8,000	0.25 m	2.5 m
	1:9,000	0.25 m	2.5 m
	1:10,000	0.25 m	2.5 m
	1:16,000	0.25 m	2.5 m
	1:20,000	0.25 m	10.0 m

Aerial Photography and Ground Truthing by TS and DGPS

It is strongly felt that this entire programme can be best done by Aerial Photography combined with ground method. A picture at the required resolution, scale and accuracy is capable of satisfying all stakeholders in terms of speed, accuracy and cost effectiveness. Today every private land owner as well as the Government for the land owned by them expects the Government to take steps to grant them with the title of the land and to reduce land disputes as part of good governance. Aerial Photography is best suited for this job. This will be the only media which will enable to satisfy the land parcel owners ultimately as cameras does not tell lies.

Aerial survey with large format digital camera at better than 10 cm GSD (Ground Sampled Distance) and automated workflow to generate digital ortho imagery is the key for creating modern land record system for India. A parcel map with topographical and Cadastral details prepared from digital ortho-image integrated with RoR (Record of Rights) is ideal to meet all the challenges in modernization of Indian land record system. Ability to map ground realities at the required scale, accuracy and precision geodetic control makes Aerial imagery a potent tool for future.

Creating a world class, transparent, efficient and integrated Land Titling System is the need of the hour today. This will help resolving land title issues and let citizens monetize the true value of their land. In rural area this would help financial services providers transparent and easy money lending to agricultural community.

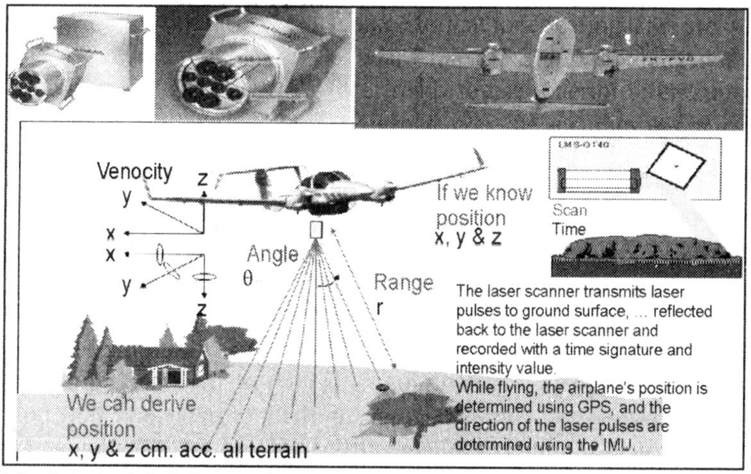

Fig. 3.4: Aerial Flying Technology

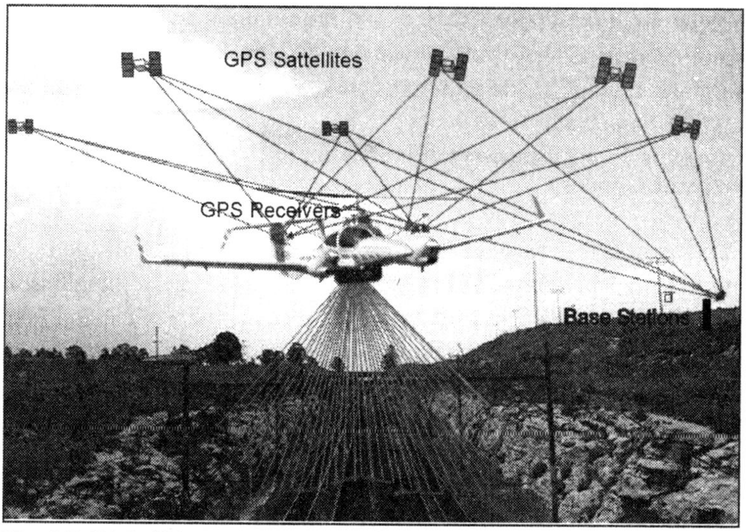

Fig. 3.5: Digital Photogrammetry (GPS Geodetic cadastral control + Image + Topographic features+map) Almost 150 sq. km survey in one hour with aircraft

There is need to have a functional Computerized Land Information System for all these things to happen. The existing cadastral data in the form of parcel maps converted into a digital format are really innumerable. Obsolete prevention techniques of the land records must be replaced by computerized updating system, which retains accurate spatial information along with attribute information.

Computerized Land Information System can minimize boundary disputes and consequently, civil litigation can efficiently safeguard the survey frame on ground. The State government can harness computerized database of every feature of land records in terms of survey fields and subdivisions for macro level and micro level planning activities to boost all important development activities. Computerized Land Information System will ensure easy and instantaneous availability of correct record to the public and thereby helps in introducing the much warranted transparency of public records in this important domain. Since all the survey field is computerized, the maps generated by the computer are, far more accurate than the maps available at present and can function as an efficient decision making tool.

It is felt that it may be a challenge for the available technologies to generate a land parcel map of the required accuracy in absolute terms of the parcel with number of kinks and bends reproduced below. Whereas this LPM has been generated from aerial photographs (Fig. 3.6).

Areas of Concern

Land Records in India has a long history which can be broadly divided into 3 periods pre-and post-British period and post-1947. This subject became a State subject in 1905, when the responsibility of maintaining land records was transferred from Survey of India to the provincial Governments. In most of the States, the land records are being maintained at Taluk/Tahsil/Anchal offices in the form of maps, sketches, registers and Field Measurement Books.

The present land records were generated using crude survey instruments such as Chain and Cross-Staff which took decades to map the land parcels. These outdated methods in use had generated large arrears due to which the maps and the land registers could never

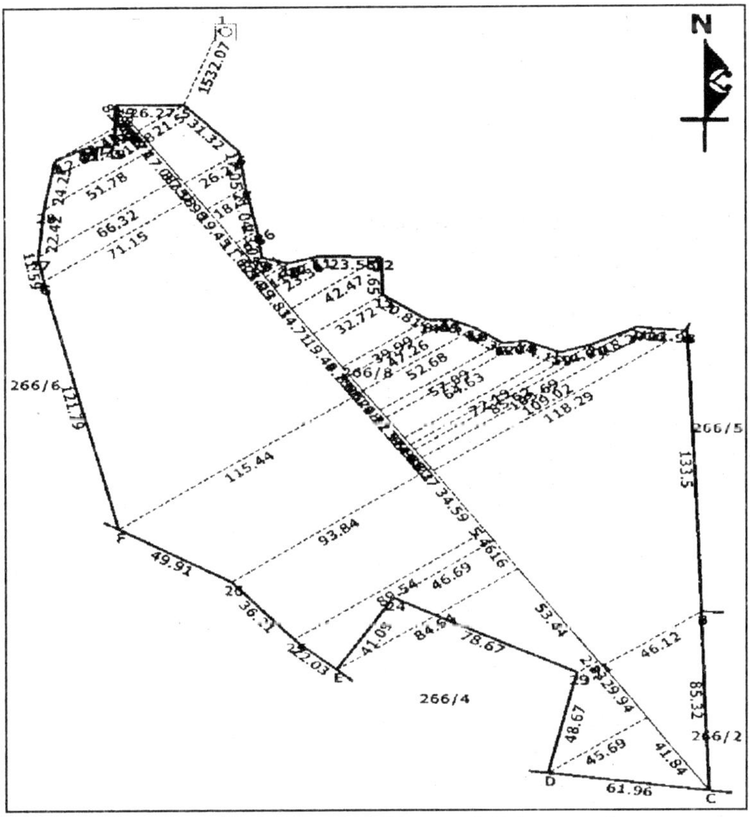

Fig. 3.6: Land Parcel Map Generated from Aerial Photograph

remain reasonably updated. Elaborate and time consuming land settlement operations also contributed to the delays in finalizing the land records.

With these ground realities, the first and foremost area of concern is to give this programme, wide publicity by different means to bring awareness at village level for the land owners both individuals and Government to demarcate their land parcel boundaries. The revenue authorities should ensure that demarcation of land parcel boundaries is done before the flying missions are undertaken. The Land Survey Act should be amended to make provision of new technologies. The most challenging area for the officials of the State Government is to get the Record of Rights register updated as this task has not been performed for ages by the land parcel owners due to lethargy and other reasons. All existing records have to be converted into digital format as very little has been done under the Government of India's similar programme of computerization launched long back. Each Taluka/Tahsil/Anchal offices have to be equipped with required hardware and software along proper trained manpower.

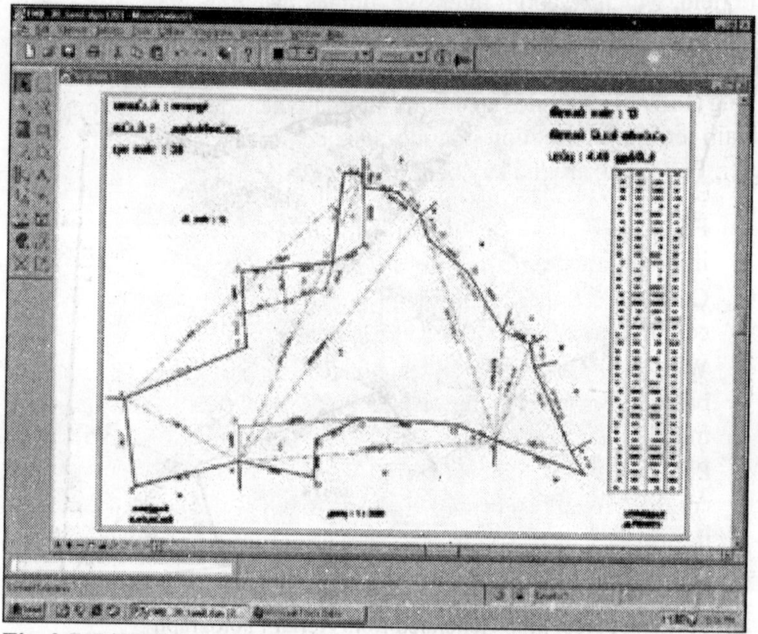

Fig. 3.7: LiDAR (Light Detection and Ranging) Technology

It is strongly felt that the third dimension, i.e., the element of height must form part of the coordinate system so as to have complete large scale maps for various uses. These maps can form the basis for compilation of topographical maps. LiDAR, in its increasing variety of operational formats, is becoming the technology of choice for fast, accurate and quality geospatial data acquisition. LiDAR technology fulfill this role of obtaining the third dimension when this sensor is flown along with aerial camera. This combination delivers the results needed for capturing geospatial data in its complete form.

It is, therefore, recommended that LiDAR Sensor may be used along with digital aerial camera to obtain high resolution spatial data to implement this Programme. The digital Aerial Image and elevation base will provide the true ground information foundation for generating other base layers and numerous thematic information by various stakeholders of Governments and local governments within the State.

Conventional surveying methods entails data collection entirely in the field, which is labour intensive, time consuming and limits of extent of information which can be obtained and digitized. The level of accuracy obtainable in Aerial image and Light Detection and Ranging (LiDAR) provides an unique alternative modern technology to obtain terrain information in a more expedient and cost effective manner. Following are the key benefits of aerial data:

- Fast and efficient way to acquire digital data over large areas in digital mode;
- Cloud free and factual ground image information for every centimetre of land without extrapolation;
- With its uniform accuracy the images can form the universal base and common framework for building all other GIS layers in the State;
- Precise geodetic control across whole State linked to local control network and automated data processing, retrieval and information analysis;
- Updating is less expensive once universal data is generated at centimeter level of accuracy using aerial photography; and
- The digital image data once acquired is useful for multiple purposes (disaster, planning, infrastructure projects).

As part of our service offering we propose to acquire the image data at less than 10 centimetre in GSD (Ground Sample Distance) and an elevation information from LiDAR for every square metre of the ground. The images with Elevation Base will provide base multiple projects such as modernization on land records, Disaster planning, Irrigation, infrastructure and rural and urban development. Following is the broad list of departments of the States which can utilize this data.

- State Planning Board;
- State Disaster Management Authority;

- Urban Development—Town and Country Planning; Municipalities;
- Irrigation Department;
- Highways and Roads;
- Flood Plain Zoning;
- Internal Security;
- Wasteland Management;
- Groundwater Board—Water Prospect Mapping;
- Forest Department—Forest data base, forest classification;
- Rural Development—District Development Offices/Zilla Panchayat; and
- Mining Department—Mine Lease and Application Processing System.

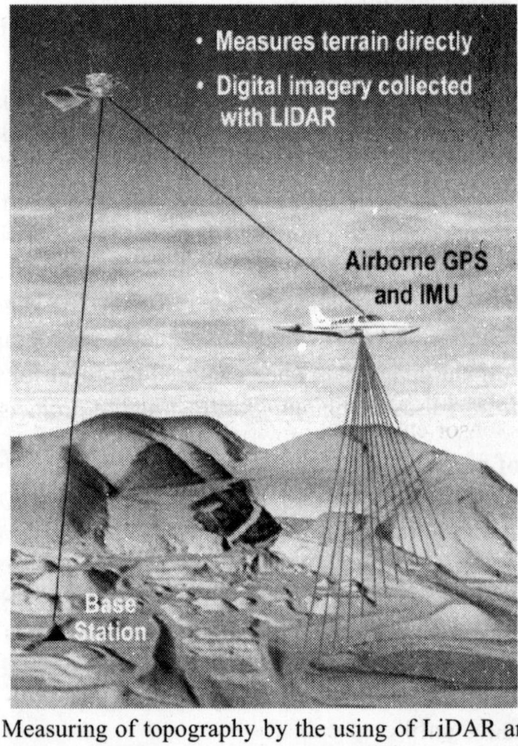

Fig. 3.8: Measuring of topography by the using of LiDAR and Airborne GPS

Applications

- Utilities
  - Electric transmission and distribution
  - Water and sewer
- Oil and Gas
- Land and Property Management
- Transportation
- Forestry
- Corridor Mapping
  - Telecommunications
  - Fibre optic routes
- Pipeline Transmission
- City, Municipal mapping
- Agriculture
- Mining
- Large Scale Mapping
- Disaster Prone Area Mapping

LiDAR Concepts

LiDAR (Light Detection and Ranging).

- Uses a laser beam to scan the atmosphere.
- The laser scanner is mounted along with an Inertial Measuring Unit (IMU) and Airborne GPS.
- The sensor emits rapid pulses of infrared laser light.
- The LiDAR system requires a surveyed ground-base location
- LiDAR scanning can occur at day or night.
- Features such as buildings and automobiles are included during scanning, are filtered by post-processing filtering techniques.
- LiDAR technology offers fast, real-time collection of 3-D points, quickly employed to generate DTMs.
- Pulse Return contains x, y, z coordinates.
- First Pulse Return—Buildings and Vegetation. Used to produce unfiltered datasets.
- Last Pulse Return—Ground below the Vegetation. Used to produce bare earth elevation data.

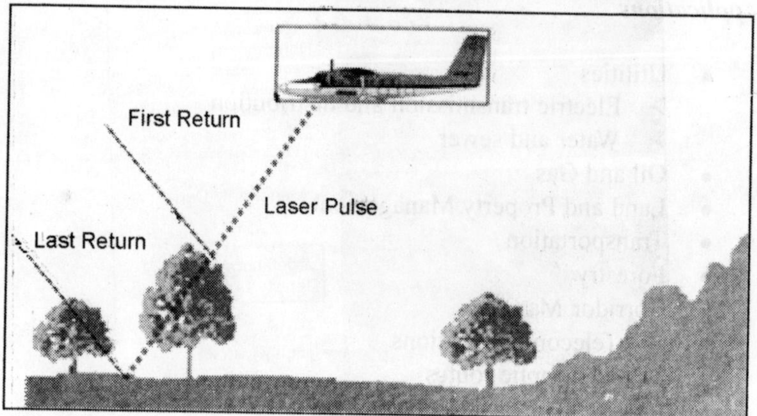

Fig. 3.9: Mechanism of Data Collection from Earth Surface by Airborne LiDAR Technique

Special processing must be done to filter out buildings, vegetation and other obstructions that do not represent bare earth.

Table 3.3 below shows the LiDAR point density statistics

Table 3.3: LiDAR Points Density Statistics

Post Spacing	*Flying Height*	*FOV Degrees*	*Swath Width*	*Pulse Area HZ*	*Project Area Per Sq. Mi.*	*Point Per Sq. Mi.*	*Total No. Points*
5 m	10,000'	55	10,411'	23,100	16,841	384,763	6,479,800,000
3.5 m	8,500'	45	6,602'	28,900	16,841	748,174	12,600,000,000
2 m	6,000'	45	4,970'	32,800	16,841	1,324,420	22,299,500,000

LiDAR Processing

Laser scanner data (point clouds) has to be processed:

- Correction of systematic errors,
- DEM extraction;
- Filtering, feature detection and extraction;
- Large volume of data generated (usually several GIGA bytes), the processing and packaging of the data has to be highly automated.

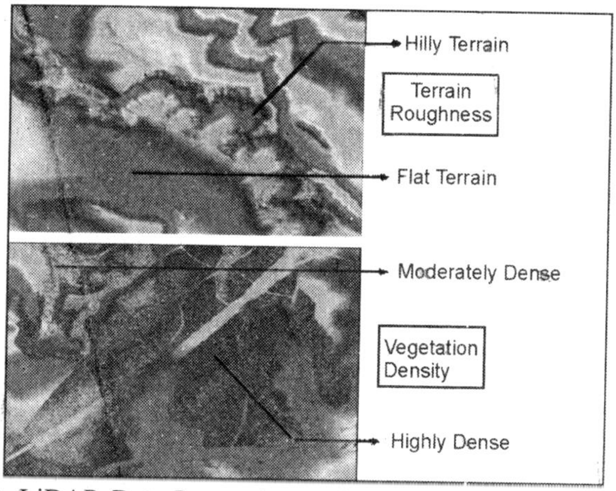

Fig. 3.10: LiDAR Data Processing and Interpretation of several Earth Features

LiDAR data processing and analysis is a function of factors such as:

- Terrain Roughness,
- Types of Earth Surface Features,
- Vegetation Density,
- Mixed Terrain and Surface Features, and
- Spatial Resolution of the LiDAR data.

LiDAR Classification

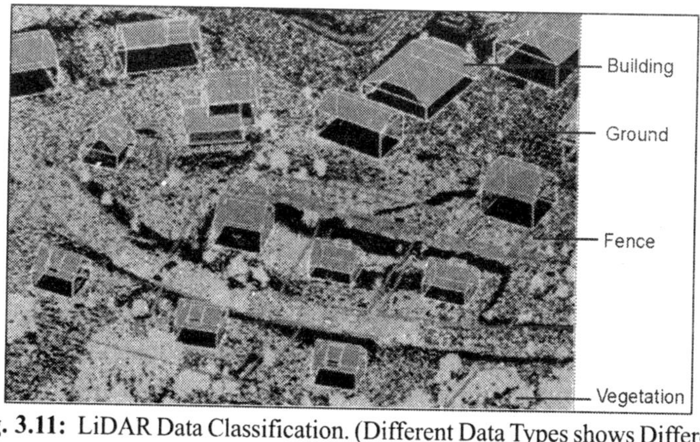

Fig. 3.11: LiDAR Data Classification. (Different Data Types shows Different Earth Features)

LiDAR Outputs with Aerial Cameras

DEM Generation

- By filtering the unwanted LiDAR points (points that are not on the ground) can be removed.
- Once the non-ground points are removed, you will be getting ground points that can be used for bare earth modelling helping in creating ground contours.
- As LiDAR gives you a very dense point cloud (more than 1 point per sq. m), contours of 0.25 m/0.5 m can be generated.

LiDAR with Aerial Cameras

Ortho Photo Generation

In some of the recent practices, aerial photo acquisition will be carried out along with LiDAR mapping. This will help in generating orthophotos for the project area very quickly.

Automatic Vectorization

As the present day LiDAR systems generate a very dense point cloud, user community is using it for feature capturing and modelling as well. Vectorization of building edges, roads, transmission lines, etc., is possible in such a high density LiDAR data set.

LiDAR Accuracy Concepts

LiDAR with aerial camera is a powerful aerial survey technology which is becoming popular in deriving Digital Elevation Models (DEM) and Digital Terrain Model (DTM) with a stunning accuracy of 10-15 cm.

- This ability of LiDAR systems to capture accurate spot heights at an extremely rapid rate is the principle reason behind LiDAR's success.
- GPS data from the aircraft and multiple ground stations are

processed together using sophisticated kinematic GPS post-processing software.

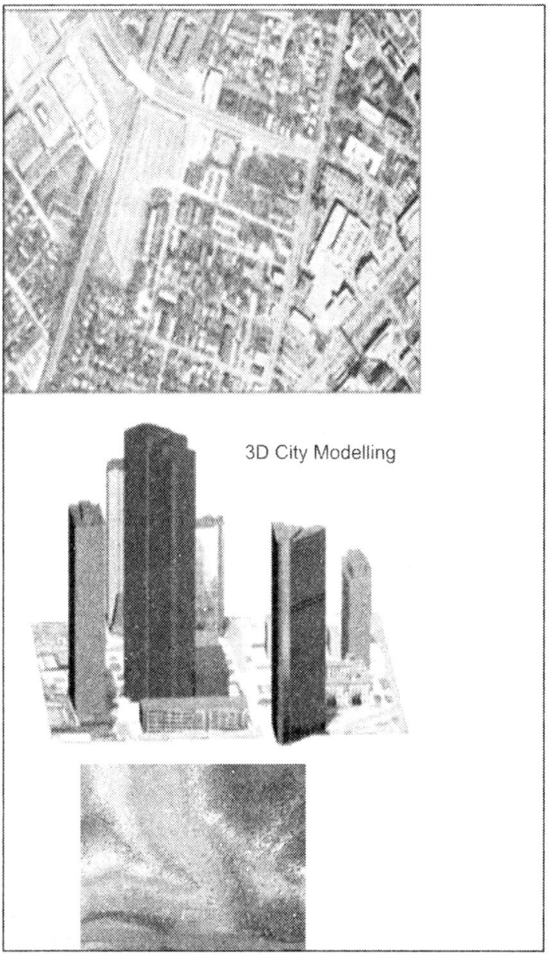

Fig. 3.12: Creation of DTM and DEM by LiDAR Print Data

- The use of two or more ground stations provides quality control and improves the accuracy of the kinematic trajectory.
- The position (x, y, z) of an airborne GPS antenna at an interval of 0.5 or one second is calculated.

Table 3.4: LiDAR with Aerial Camera— Accuracy Concepts

Point Spacing	*Vertical Accuracy*	*Contour Interval*	*Application Supported*				
			Base Mapping	*Flood Plain Mapping*	*Natural Re-sources*	*Transpor-tation/ Utility*	*Urban Modelling*
1 m	0.09 m	0.33 m	✓	✓	✓	✓	✓
2 m	0.20 m	0.66 m	✓	✓	✓	✓	
3 m	0.30 m	0.99 m	✓	✓	✓		
4 m	0.04 m	1.00 m	✓				
<1 m	0.20 m	0.25 m	✓				

Flying Height	*Horizontal Accuracy*	*Vertical Accuracy*
20,000 + ft.	± 1.5 m (4.9 ft.)	± 0.6 m (2.0 ft.)
8,000 ft –20,000 ft.	± 0.75 m (2.4 ft.)	± 0.3 m (1.0 ft.)
5,000 ft –8,000 ft.	± 0.4 m (1.3 ft.)	± 0.15 m (0.50 ft.)

Table 3.5: Comparison between LiDAR and Aerial Photogrammetry

Features	*LiDAR*	*Photogrammetry*
Energy source	Active	Passive
Geometry	Polar	Perspective
Sensor type	Point	Frame or Linear scanning
Point measurement	Direct	Indirect
Sampling	Individual points	Full area
Associated image	None or monochrome	High quality spatial and radiometric
Horizontal accuracy	2-5 times less than vertical accuracy	1/3 better than vertical accuracy
Vertical accuracy	10-15 cm (10 cm per 1,000 m over heights of 2,500 m)	Function of flying height and focal length of camera
Flight planning	More complex due to small strips and potential data voids	Overlap and side lap need to be considered
Flight restrictions	Less impact from weather, day/night, season, cloud condition	Must fly during day and need clear sky
Production rate	Can be more automated and faster	Can be automated in data processing
Budget	25-33% of Photogrammetric compilation budget	More or less equal to satellite Stereo
Production	Proprietary software: processing performed by vendors, operators	Desktop software available to end-user
Limited contrast area acquisition	Can acquire data: used extensively for—coastal mapping/Flood/ Disaster	Medium expensive

Web based Land Information System

The land parcel data once created by Aerial Survey having both the static and dynamic elements needing updating automatically or at regular intervals manually. This database will be put on the web duly customized to offer multiple query based facilities and its updating by the concerned department of the Government. This web based parcel-wise land information system will have the facility for extraction of information about the ownership and other details. The multi query facility will allow together information regarding landholding.

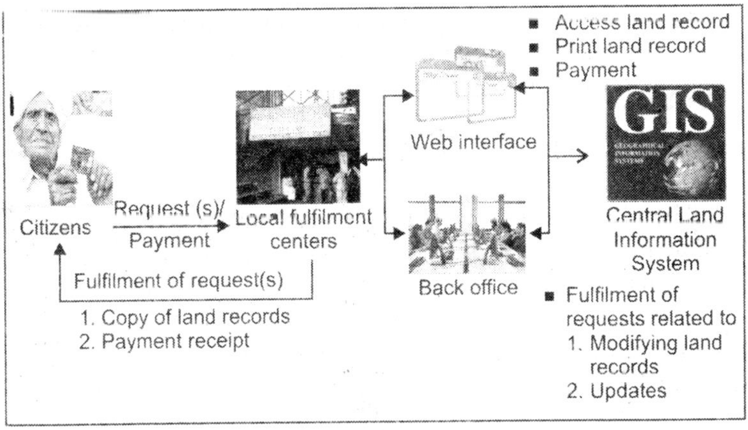

Fig. 3.13: Citizens Accessing Land Information Systems

This web based LIS will create a world class, transparent and efficient Land Records System to meet the following requirements of various stakeholders:

- A single window to handle land records transactions.
- Cadastral records should mirror the ground reality.
- The record of title is a true depiction of the ownership status.
- Automated mutation during land transaction.
- Open up opportunity for Title Insurance Companies to start their operations to guarantee the title for its correctness and

indemnifies the title holder against loss arising on account of any defect therein.

- Loans to the Agricultural community will become easier.
- Allow citizens to monetize the true value of their land.
- Will create thousands of jobs across the States.
- Help Government bring in transparency in Land Records Maintenance.
- Create systems which will allow access to the database as well as allow mutation of the records in a secure transparent manner.
- Easy tracking, maintenance, and linkages with previous transaction.
- Make the process of registering land transaction simpler.
- Create Access points where citizens can access this information by making nominal payment.

Conclusions and Recommendations

It will be advisable to execute this National Land Record Management Programme to make use of the technology of Aerial Survey with LiDAR Sensor data. The choice of this technology will give the States the required scale, accuracy and speed with additional benefits of having extensive data pertaining to the third dimension — the height, required for various other purposes. In addition to the advantages described above Web based Computerized Land Information System so generated will have the following salient features.

- Efficient visualization of GIS data with high resolution aerial Ortho Photos in backdrop.
- Enhancing mutation workflow by putting industry standard authentication, approval and security system.
- Backdrop Aerial images with terrain information helps in efficient mutation process.
- Automatic generation of digital land parcels from Field Measurement Book (FMB)/Tippan tables, where updated cadastral maps are not available.
- Title history management for future reference and RoR purpose.

- Showing property images in urban areas for proper taxation and encroachment detection.
- Linkage between GIS data with land parcel data and relevant documentation using unique coding mechanism.
- Real-time parcel updating and information flowing to central data centres to minimize post processing work.
- Functionalities to be supported by fulfilment centre software such as full-fledged land parcel editing tools, scan and upload of reference documents, biometric authentication system, etc.
- Provide auditing system, edit logs, undo options and versioning.

ANNEXURE - I

Comparison of Three Survey Technologies

Details	Ground Method	Aerial Photogrammetry	HRSI
Suitability	For small areas	For large areas	For large areas
Terrain	Suitable in plain areas	Suitable in all areas	Plain and Hilly areas (Only with stereo data)
Ground Control	Extensive	Limited	Limited
Data Acquisition	One Survey Team can complete an out-turn of 0.25 sq km per day	500 sq km can be flown in a day.	Depending on the Satellite revisit coverage and weather condition.
Manpower Requirement	Large number of trained and experienced field oriented Surveyors	Limited number of Technology professionals	Limited number of Remote Sensing professionals
Costing	Expensive Field Work	Cost effective	Cost effective
Accuracies 20cm/10cm/5cm for Rural/ Semirural/Urban	Achievable as laid down in Land Survey Act with precise instruments like TS and DGPS	Achievable as per Land Survey Act with limited use of Ground truth	Barely achievable with extensive Ground control network support
Level of Details	Limited	As small as 2.5 cm (Every Detail)	Around 50 cm (Topographic Details)

(Contd.)

Annexure - I (*Contd.*)

Details	*Ground Method*	*Aerial Photogrammetry*	*HRSI*
Scale of Mapping	Cadastral Scale	Suitable for Cadastral and Topographical mapping	Suitable mainly for Topographical mapping
Utilization	Cadastral Survey. Impractical for Topographical and Polygon Survey.	Multipurpose (Topographic + Cadastral)	Multipurpose (Topographic)
Availability of data	As per requirement	As and when required	Depends on the weather and cloud cover
Change Detection and Updating	Time consuming	Possible	Possible
Instruments	Large number of precise field instruments	Very limited number of Instruments	Very limited number of Instruments

Chapter – 4

NLRMP—Implementation Issues and Challenges

Vinay Thakur, D.S. Venkatesh, D.C. Misra, Surendra Kumar and *Charanjit Singh*

Introduction

Land plays a crucial role in our social and economic life, particularly in rural areas. The social and agrarian fabric could be disturbed if land management issues are not adequately addressed. The accuracy of basic land records data will enable the Government to scientifically plan development programmes for the society and help in good governance. Land is partitioned for administrative and economic purposes and it is used and transformed in a myriad ways. Population growth, technological, social hazards and environmental degradation have all to be taken into greater account today by policy makers, resource planners and administrators who make decisions about the land. Identification of beneficiaries for Government schemes is easily identifiable thereby reducing inordinate delays and lead to best practices reducing corruption. More recently, the need for thoughtful and careful stewardship of the land, together with the more intensive use and management of its resources has emerged as a matter of major global concern. Knowledge base and business intelligence can be built wherein the other land related parameters such as crop yield, land-use pattern, soil, size of holdings, source of irrigation data can use the accurate basic land records data. Although the printed map is still useful, computerized systems offer improved ways of acquiring, storing, processing and retrieving critical information about land.

Computerisation of Land Records (CLR)

The scheme of computerisation of land records was started in 1988-89 with the aim of removing inherent flaws in the existing land records system. It was a centrally sponsored scheme with 100 per cent financial assistance from Government of India. The main aim of the scheme was to computerise ownership and plot-wise details for issuance of timely and accurate copy of the computerised Record of Rights (RoRs) to the land owners from tehsil computer centre from an estimated 4000 Tehsils/Taluks. Some of the States/UTs have also stopped the issuance of hand written RoRs. There are fourteen States, which have provided web interface for citizens to view their RoR. The basic work of converting the Record of Rights (RoRs) into digital form has been completed in majority of the States. The other deliverables of CLR included providing Land Records services to the citizens through Kiosks (both coinbased and touchbased) and automated workflow in the Tehsil/Taluk/Revenue Circle level. Bhoomi of Karnataka, Tamil Nilam of Tamilnadu, Himbhoomi of Himachal Pradesh, HALRIS of Haryana, Bhumi of West Bengal, Bhulekh of Uttar Pradesh and Bhuabhilekh of Madhya Pradesh, Apna Khata of Rajasthan, Mahabhulekh of Maharashtra, Jami of Tripura, Dharitree of Assam, Dharini of Goa are some of the Land Records application software from NIC which are very popular in the respective States and have earned wide laurels. These products have also been ISO tested for quality.

Other output documents such as mutation extract, khata register, generation of mutation notices, generation of new jamabandi, various other reports, demand and collection, etc., are also some of the other deliverables of CLR. Security of land records data, usage of biometrics (finger print) authentication, digital signature, normalized database and ISO Quality certification are some of the salient features of the application software.

Strengthening of Revenue Administration and Updating of Land Records (SRA and ULR)

Strengthening of Revenue Administration and Updating of Land Records was started during 1988-89 with 50 : 50 sharing basis between the Centre and the State. Updation is vital for systematic maintenance of land records to reflect ground realities in sync with ownership changes, ensure genuine land transactions and implement rural development programmes for effective enforcement of land reforms. It will minimize land disputes and create social harmony in villages. The main objectives of the schemes are:

- Strengthening of survey and settlement organizations for an early completion and preparation of land records in areas where this work still remains to be done.
- Setting up of survey and settlement organization especially in the northern regions whrre no land records exist.
- Imparting the pre-service and in-service training of revenue, survey and settlement staff and strengthening of training infrastructure in their purpose.
- Providing the facilities for the modernization of survey and settlement operations, printing of survey maps, reports/ documents and for storage, copying and updating of land and crops records using, amounting other things, science and technology inputs.
- Strengthening of revenue machinery in the village and immediate supervisory levels on a selective basis to make the workload of these functionaries manageable.

Computerisation of Registration Process

This was not covered in either CLR or SRA and ULR earlier. Many States had taken initiatives to computerize different processes of registration using their own funds or outsourcing some processes in lieu of collecting user charges. A few pilot projects were sanctioned by Department of Information Technology (DIT), Government of India.

Registration means recording of the contents of a document with a Registering officer and preservation of copies of the original document. The documents are compulsorily registered for the purpose of conservation of evidence, assurance of title, publicity of documents and prevention of fraud. The main function of the Sub-Registrar offices (SROs) is to register documents under the rules and regulations of the Indian Registration Act 1908. Documents are classified according to the nature of the transaction. Registration is mandatory only for certain types of documents. The importance of registration is as follows:

- Stop fraud cases of registration.
- To give solemnity of form and legal importance to documents.
- Perpetuate documents, which may be of legal importance in future.
- Give complete assurance regarding the validity of the document.
- Make available a copy of the document at any later point of time. For this purpose the office maintains a photocopy of each and every documents registered by the office.
- Preparation of Index 2 for issue to the concerned parties as proof of title.

Need for Computerisation of Registration Offices

The work-culture and age-old procedures have brought following problems in the office of Sub-Registrar:

- Lack of high degree of transparency in valuation of properties and calculations of stamp duty and registration.
- The system is not user-friendly to fulfil the desires of public.
- Delays in providing the services and office record maintenance.
- Problems in voluminous document management.

Above issues called for radical reforms so that rigid and complex system can be simplified to bring the transparency and one-stop service to the common man. The registration system implemented in States is able to automate above processes and bring transparency and efficiency in entire system.

An estimated 2900 Sub-Registrar offices are computerised. CARD of Andhra Pradesh, PRISM of Punjab, HARIS of Haryana, RED of Gujarat, CORD of West Bengal, SARITA of Maharashtra, KAVERI of Karnataka, HIMRIS of Himachal Pradesh, STAR of

Tamilnadu, PEARL of Kerala, PRERNA of Uttar Pradesh are some of the success stories of Computerisation of Registration process by NIC. These States including Tripura, Delhi, and Puducherry (UT) have achieved different degree of success in computerising registration of immovable property transactions.

Survey and Cadastral Mapping

Traditional methods of generating cadastral maps are from the usage of conventional instruments such as chains and tapes which were not only time consuming but error prone. The latest technologies used in survey today are Electronic Total Station (ETS) and Global positioning system (GPS) which when used together are able to generate accurate cadastal maps. Aerial photography and high resolution satellite imagery are also used in today's scenario.

Cadastral maps are related with parcel boundary and in scales like 1 : 2000 or 1 : 4000 and are for a smaller area like a village. The topographical maps are of very high scale (1 : 50000 or 1 : 250000). The topographical maps provide details for large areas with emphasis on broad natural features and generally prepared by Survey of India (SoI) and the maps are classified as public and defence series.

The objectives include preservation of old and worn out maps and updation of land records. There is also a need for modernization of survey operations and need for integrated system.

Organization of Cadastral Data

There are two ways in which spatial data has been organized. Village maps exist in an estimated twenty States spanning north, eastern and north-eastern States. In the south of Vindhyas, the spatial data on individual land parcel in the form of tippan or Field measurement book (FMB) is maintained. Tippan are created from field measurement data and exist in an estimated seven southern States and southern and western States/UTs. Variations do exist in both the systems.

Implementation Strategy for Digitization

In States/UTs which are having the ladder data as the basic records, the tippan has to be created from the ladder data and used for updation. The tippans can then be mosaiced to generate village map/taluk/tehsil maps, etc., for planning and seamless integration.

In States/UTs which are having the cadastral village map as the basic records, the village maps have to be digitized and these sheets need to be mosaiced to make village maps.

Accuracy Requirements

The existing maps have an inherent error of an estimated 2.5 per cent. Care should be taken to ensure no error be introduced during digitization. Errors could creep in due to shrinkage of maps and also because of skew maps. Thickness of lines could also introduce errors. The errors can be avoided if rubber sheeting is done carefully; a higher scale such as 1 : 1000 or 1 : 2000 must be used; keep dots per inch (dpi) as high as possible. In any case, errors due to digitization should not be more than 0.025 per cent. Revenue staff should validate 100 per cent by edge matching of digitized printouts with original maps through superimposition. In southern States/UTs, individual cadastral maps can generate accurate village maps. While doing so, it will be seen that the mosaic maps will spill over the village boundaries. This can be fixed carefully by using optimization techniques.

Some Cadastral Map Data Standards

- Shape file (shp)
- Integration with GML3.0
- Land XML support
- Support for linkage with Aerial Photography
- Satellite Imagery Integration
- Provision of data input from GPS/ETS and other hand held devices.

Scanning Process

- Source document will be paper/cloth bound cadastral maps normally in A0 size on the scale of 1 : 3600, 1 : 3960 and 1 : 4000. This source document will be scanned at land records office only. The sheet has to be cleaned. The vendor has to scan at not less than 300 dpi and format must be in tiff format (*. tiff).

- After scanning, one copy of map on 90 micro transparencies will be printed and cross-checked with the original input map given by respective tehsils. The vendor has to make sure that original document and printed document are 100 per cent matching. Output will be submitted as **scanned image on CD** and hardcopy **on good quality GSM paper.**

Issues and Constraints in Land Management System

Land Records

Process Related

- The land records data is available in registers which are with the revenue functionaries.
- Data is maintained in local language of the States/UTs as land related matters are with the revenue department which is a State subject.
- Computerisation cannot be made operational unless the entire database is initially created, validated, verified and corrected.
- Resistance of the lower level revenue functionaries to get involved in the project is another aspect.
- Land related documents have legal implications and hence cannot be changed and hence are retained in the computerised system. Hence, no business process re-engineering (BPR) has been attempted.
- The end users are used to the old formats, especially the source documents, hence the input screens and output reports cannot be different.
- High volume of data is an integral part of land records system.
- Legacy data needs to be preserved as the courts require land records data at any given point of time. Hence, the mutation process and other processes have data redundancy to a large extent.

- Different local or colloquial terms being used to record the characteristics of land, soil, land-use, etc.
- There is no standardization of operating procedures to generate textual and graphical records.
- Different practices being followed in a locality are treated as standard.
- Different practices introduced by individual village functionary such as Patwari and already recorded in the registers for the last few decades.
- Variation in area units, practice of recording of data by the village functionaries.

ICT Infrastructure

- Hardware, system software, installation, site preparation including AC, furniture is required.
- AMC is required for maintenance of the ICT infrastructure.
- Training is required to ensure minimum downtime of the system.
- Proper authentication measures such as password, biometrics (finger print readers), digital signature for providing secure access to authorised personnel only.
- When data is available on the website for viewing purpose to the public, care has to be taken to ensure 24 × 7 uptime with proper disaster recovery for any eventuality.

Others

- Parallel operation of both the computerised and manual system.
- Non-stoppage of hand written RoRs
- Non-availability of the land records data on the website.
- Training and Capacity building are a continuous process owing to the frequent transfer of the revenue functionaries.
- Deployment of revenue staff and officers for flood relief work, drought relief work, census operation and other non-revenue related works is also a constraint.

- Mutations are not initiated by the revenue functionaries on a *suo motu* basis.

Property Registration

Process Related

- Guideline values—only with SUB-REGISTRARS.
- Adjudication of stamp duty—prerogative of SUB-REGISTRARS.
- Encumbrance Certificate—abnormal delay due to manual search, inadvertent omissions.
- Writing of document—Document Writers' menace as middlemen.
- Copy of document only by Licensed Copy Writers/Typists.
- Registration process—manual endorsements on copies.
- Certified Copies of documents—to be written manually by document writers.
- Preservation of Records—huge volumes requiring space.
- Registration of Marriages—manual search for issue of certificate.
- Registration of societies/chits/partnerships—poor retrieval of information and tracking of activities.
- Sub-registrars are not sending registration details.
- Title is not verified at the time of registration.
- Buyers are not filing request to update ownership details.

ICT Infrastructure Related

- Hardware, system software, installation, site preparation including AC, furniture is required.
- AMC is required for maintenance of the ICT infrastructure.

Others

- Key resource persons are identified and given intensive training on the application of software.
- Regular training is also imparted to the SRO staff and officials on the usage of the system.

Survey and Cadastral Mapping

Process Related

- Conventional methods of survey include chain and tapes are time consuming and error prone.
- Total Stations are accurate but again consume more time.
- Pilots tried earlier yielded mixed results.
- Old and worn out cadastral maps.
- Absence of cadastral maps for some land parcels.
- Updation of land records and maps.

Localization Issues

- Local language support is necessary for Land Records application.
- Land Records data entered in DOS/Foxplus environment had to be migrated to Unix/ORACLE and later on Windows/SQL Server without any data loss and at the same time, the local language flavour had to be maintained.
- Migration utility had to be developed whenever there was change in platform so as to maintain the legacy land records data.
- Usage of GIST SDK for storage of data in both 7/8 bit ISCII and also in ISFOC, although rendering of output was in ISFOC with suitable bi-lingual fonts.
- At the same time, when the data had to be displayed on the website, at first, suitable fonts had to be downloaded in order to view them.
- As an attempt towards standardization of storage as Unicode, Some States have migrated their data into Unicode.
- The original intent of Unicode was to represent only the basic linguistic elements forming the alphabet and not a specific rendered form.
- In Unicode, words have to be interpreted properly to extract linguistic information and this requires analyzing the syllable structure.
- The data types for fields in the database table have to be

converted to nvarchar data type and char to nchar data type as most of the States have their data in SQL Server.

Standardization Issues

- Land Records data are maintained in various States following State specific set of codes for different attributes such as crop, soil, season, land-use, source of irrigation, area units and location.
- There is no uniform list for these attributes which describe the plot.
- Even if the list is there, it is incomplete and not uniform throughout the State.
- Different colloquial/local terms are used for indicating characteristics of a plot.
- Different practices introduced by the individual village revenue functionary and already recorded in the registers for the last few decades.
- Different practices followed over the decades in a locality are treated as standard.
- Non-adoption of any scientific methodology to name the characteristics of the plot.

Need for Interoperability

- Fifteen parameters identified at the national level for standardization.
- There is a need for standardization at the national level for data warehousing and data mining purposes.
- This will facilitate multipurpose information and knowledge exchange within Land Records domain and also collaboration with other national level databases.
- Standardization will facilitate collaboration and interoperability.
- National level data repository will help for planning and management information systems.
- Look up table have been created (transformation approach) so as not to disturb the existing system practiced.

- This will lead to developing metadata standards for land records domain.

Issues which Need to be Addressed for Data Exchange

- Mapping the SR Offices with Tehsil/Taluk offices.
- Standardizing the data exchange format/Schema.
- Suitable security and authentication mechanism for ensuring secure access and exchange of data between authentic sources.
- Amending the acts and rules to facilitate exchange and interoperability.

Challenges in Land Management System

- Computerization in silos.
- Actual benefit of computerization is not possible without tight integration between all three wings namely Revenue, Registration and Survey.
- Perfection in terms of procedures, data and citizen friendliness can only be brought in by coordinated efforts of all three wings.
- All three wings should try to converge at single vision of computerization to achieve Total Land Management System.
- Skill upgradation of the officials and field staff by having capacity building programmes regularly.
- Identification of master trainers as nodal officers for effective implementation.
- Legal changes to move from presumptive titling to conclusive titling as envisaged in the guidelines of NLRMP.
- Replacement of old and obsolete hardware in tune with the latest State of art technologies.
- Focus was more on computerisation of records and less on updating.
- No integration of textual and spatial data.
- Cadastral maps and records in bad shape, outdated, or even non-existent (ex., in some North-Eastern States).
- Survey/Resurvey using traditional methods is costly, time-consuming and error-prone.

- No linkage of land records with credit sources such as banks, insurance companies, rural and co-operative banks, etc.
- Increased security issues.
- Preservation of old and worn out maps.
- Updation of Land Records—Changes in attribute data are happening regularly whereas maps are not getting updated.
- In some States/UTs, although land records maintenance is with revenue departments, registration is with commercial tax department, law and justice department, home department while cadastral maps are with survey department.
- In some States such as Haryana, Himachal Pradesh, Punjab the Tehsildar is also the Sub-Registrar.
- Mapping of SROs with the Tehsils/Taluks/Revenue circles and connectivity among the offices to ensure seamless integration.

National Land Records Modernization Programme (NLRMP)

The National Land Records Modernization Programme (NLRMP) was launched by Government of India in August 2008, aimed to modernize management of land records, enhance transparency in the land records maintenance system, minimize scope of land/property disputes, and facilitate moving eventually towards guaranteed conclusive titles to immovable properties in the country. The major components of the programme are computerization of all land records including mutations, digitization of cadastral maps and integration of textual and spatial data, survey/resurvey and updation of all survey and settlement records including creation of latest cadastral records wherever necessary, computerization of registration and its integration with the land records maintenance system, development of core (GIS) for seamless integration and capacity building. Till date 240 districts have been sanctioned and actual implementation has started in these districts.

Need for Integration of Land Records, Registration and Cadastral Mapping

The increasing volume of transactions has led to inordinate delays and inefficiencies. Therefore, there is a need to computerize

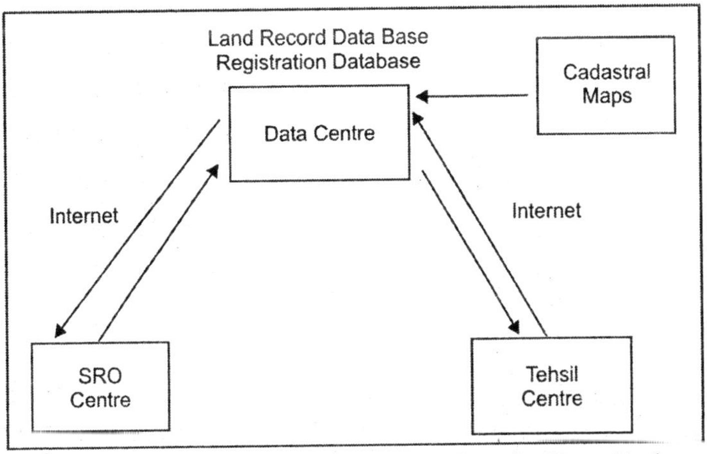

Fig. 4.1: Typical Diagram Indicating Procedure for Data Exchange/ Integration

registration process and integrate with the computerization of land records in order to streamline the process of mutation, regular updation of land records, enforce genuine transactions and better planning of land-use. This will also enable transparency and efficiency in the system.

Objectives of Integration

(a) Forward and backward linkages in terms of sharing of information of land records and registration will ensure the speedy and genuine land transactions along with continuous land records updation.
(b) Reduce scope for manipulation of records and under valuation of land and property.
(c) Reducing the time required for completing the registration and mutation.
(d) Computerize books/registers will ensure proper control and monitoring.
(e) Assist the revenue functionary in initiating mutation process for speedier disposal of mutation cases and updating of RoRs.

The registration system can use the information on ownership from land records database like viewing of the present RoR which

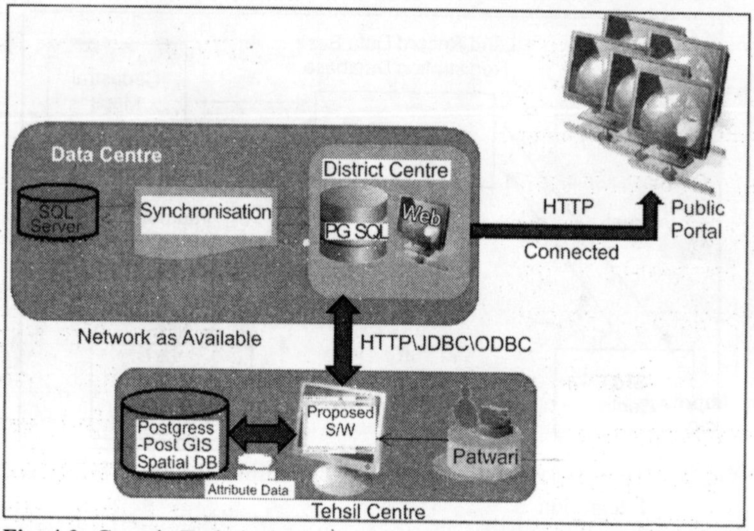

Fig. 4.2: Generic Technical Architecture of NLRMP Infrastructure

will help in verifying the deed details and also the list of prohibited survey numbers or plot ids (Khasras) must be available to the SRO to enable him know about the liabilities or charges on the land. This will help the registration system in registering genuine transactions The details about the transactions sent by the SRO to the Tehsil office in prescribed format will not only help in finding erroneous and incorrect transactions but also for initiating mutation notices and processes. By linking with the cadastral map, the RoR can be given to the land owners along with the copy of the cadastral map as a service to the citizens.

The details of registered documents must be generated and sent to the Tehsil/Taluk office daily in the prescribed format 4.1.

Integration with Land Records Required Connectivity if Not Co-Terminous

Registration of a deed is the begining point that triggers the process of mutation. Mutation notice can be generated immediately after the registration of ducument (Fig. 4.3).

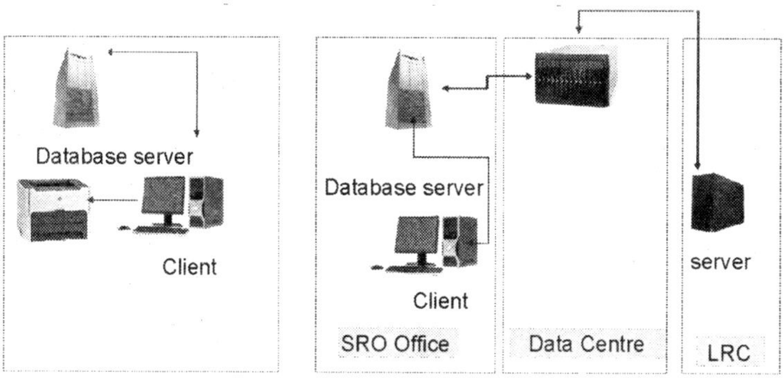

Approach 1: Store and Forward every day

Approach 2: Online Updating Land Record Database (Dedicated Network)

Fig. 4.3: Technical archtitecture for Integration of Land Records and SRO where both of them are not co-terminus

Format 4.1: Details of Registered Documents

Land Records	*Property Registration*
Location Code	Location Code
Khasra Number (Survey number, Parcel number, Plot number)	Khasra Number
Khatha Number (account number)	Area units
Area units (Extents)	Transaction type (Sale, Gift, Will, Inheritance, Pledge, Decree)
Share	Share
Owner Details	Parties Details
Crop Details	Stamp duty
Soil	Consideration amount
Source of Irrigation	Execution Date
Mutation Transaction type	Identity proof
Remarks	Land type
Land-use	Book number

Common Information in both the systems for information exchange are:

Location Code (State code, District code, Tehsil/SRO code, Village code)
Area units
Share
Parties Details (Seller and Purchaser)
Transaction type
Khasra number (Survey number)

Land Records data can also be shared with other institutions and departments such as banks, agriculture, forest rights under tribal affairs, etc., NIC has also developed a MIS for forest rights system under Ministry of Tribal Affairs. (http://forestrights.nic.in) where land patta distributed will be integrated in Land Records database.

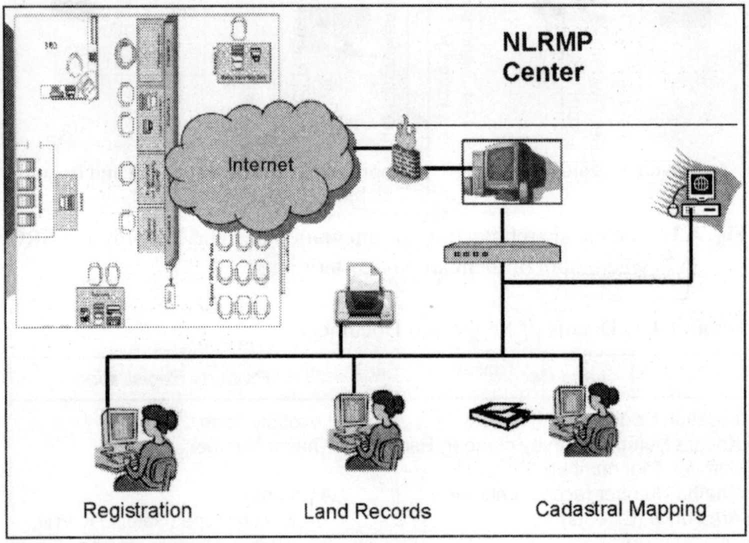

Fig. 4.4: Indicative Diagram of a Typical Citizencenric NLRMP Delivery Centre

Social Impact

- Easy access to RoR and quicker delivery. In some States/ UTs, Land Records data, Guideline values, Deed formats are accessible to the public through internet.
- RoRs can also be viewed through Kiosks in some States. Earlier, it took a long time to acquire a copy of the RoR as it depended on the availability of the copier, Patwari, Tehsildar, etc.
- Increase of collection and revenue as owing to easy accessibility of RoR, large number of public are taking their copies of the RoR for bank loans, crop insurance, etc.
- Increase in public awareness as a large section of people, that is government functionaries, empanelled private agencies, kiosk operators are involved generating employment.

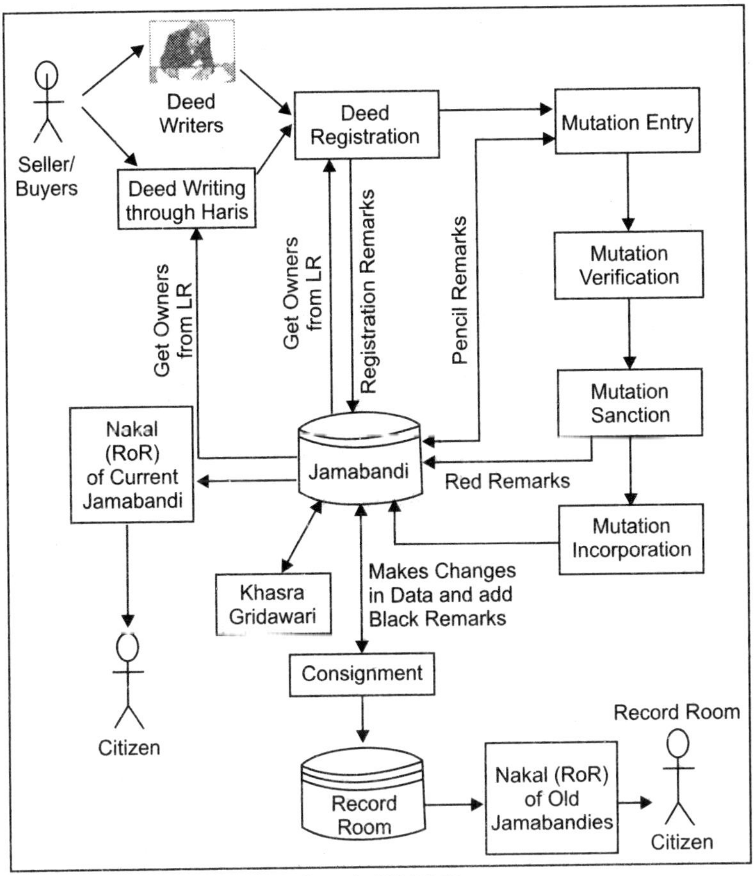

Fig. 4.5: Land Records Workflow in NLRMP

- Land Records have become transparent leading to good governance. Government or public can be alerted immediately if any encroachment is made on Government land.
- Increasing credit flow as the RoR is a vital and legal document, to acquire financial loan, this has helped the flow of credit fund to the households contributing to the growth of economy.
- After computerisation, owing to the streamlining of land records management, there has been a decline in land related disputes. The integration with SRO and cadastral maps has further resulted in reduction of litigation.

- Quicker mutation process and registration of deeds ensuring faster updation.
- Linkage with other institutions in many States/UTs, land records data have been made available on the internet and accessed by various institutions such as judicial, financial, development departments, NGOs, etc., in their respective domains.
- Creation of jobs for local youths as the project have been implemented in PPP model involving private partnership in data entry, delivery of RoR through kiosks, etc., creating job opportunities for local youths.

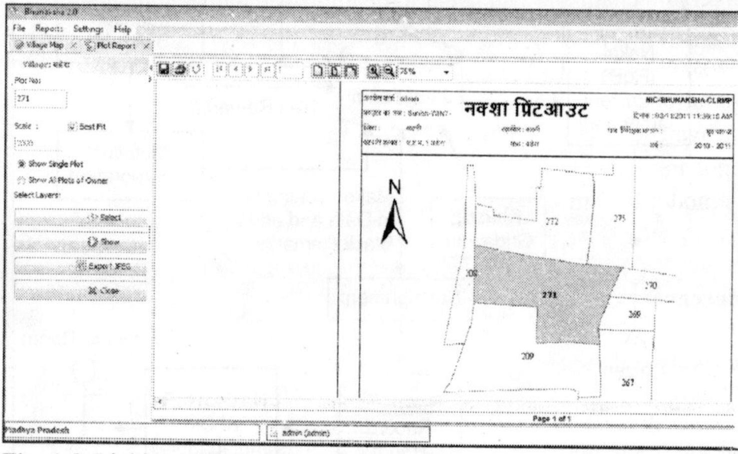

Fig. 4.6: Linking of RoR with Cadastral Map

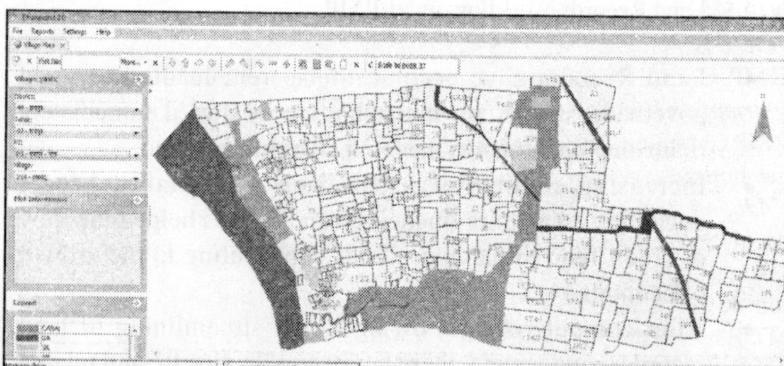

Fig. 4.7: Sample Thematic Representation on a Cadastral Map

Conclusion

Computerisation has been instrumental in implementing better land management system. People have been empowered with land records data and information. Land Records data have to be linked with SRO and the cadastral map thereby ensuring constant updation to make it real time. If the Land records data which are so critical and crucial is constantly updated, interlinking of this data with spatial data will ensure that this data can be shared with other departments and institutions thereby making it interoperable. The infrastructure and delivery mechanisms also needs to be improved with adequate power backup, internet connectivity, additional hardware, 24×7 support groups, etc. Use of advanced technology by limited automation and integration with cadastral map and registration process would further improve the revenue administration and ensure the effective and proper implementation of NLRMP a reality so that India can be a role model for other developing countries to follow.

References

Bhunaksha, cadastral mapping solution from NIC, SRS document, 2009-10.

Computerisation of Land Records: National Perspective; SD Meena *et al. Land Reforms in India*, Sage Publication, 2005.

Land Records Information System, Software Requirements Specification, NIC-LRISD, 1994-95.

LIS (2000); Vinay Thakur, Ganesh Khadenga, D.S. Venkatesh, Dr. D.R. Shukla.

NLRMP Capacity building training materials, 2009, 2011. (*http://nlrmpportal.nic.in*).

NLRMP Guidelines (http://dolr.nic.in).

Social Impact of Computerisation of Land Records, Vinay Thakur *et al. Adapting e-Governance*.

Study of Uniform Coding Scheme for Computerisation of Land Records, August 2008.

Vision document for The Computerisation of Land Records in India— D. Bhalla, Director (LR), DOLR-1999.

Chapter – 5

Land Records Data Integration Strategies

D.S. Venkatesh, Sanjukta Pradhan, Omlata, Vinay Thakur and *D.C. Misra*

Introduction

Computerisation of Land Records (CLR) is 100 per cent centrally sponsored scheme of Government of India which is being successfully implemented in more than five hundred fifty districts (in more than 4000 Tehsils/Taluks) as joint venture project of Department of Land Resources (DoLR), National Informatics Centre (NIC) and State Governments.[1, 2, 3, 4, 7]

The States' Land Records Information Systems have been customized, ***within the aegis of the core BHULEKH software,*** to encompass all specificities of the related regional and practiced land records system. Regional languages have been carefully embedded within these systems keeping in mind final integration at national level.

This project has won national and international recognition in the e-governance domain. Bhoomi project of Karnataka was awarded with international awards for its success. Similarly Apnakhata in Rajasthan; Tamil NILAM in Tamilnadu; Bhuabhilekh in Madhya Pradesh, Bhulekh in Orissa, HARIS in Haryana, LRMIS in Andhra Pradesh, Bhurekha in Kerala, Bhuiyan in Chhattisgarh, Dharitree in Assam, Jami in Tripura, Bhulekhsoft in Gujarat, Bhulekh in Uttar Pradesh, Nilamagal in Puducherry, Dharini in Goa; Bhumi in West Bengal are some of the success stories in Land Records domain.

Registration Project, has been successfully implemented by different States in an estimated 2900 SROs known as CARD, PRISM, Dastavej, ORIS, GARVI, STAR, PEARL, HIMRIS and CORD, etc.[5]

Issues and Constraints

(a) There is no uniform codification list for the land related parameters, which describes the plot attributes like soil, crops grown, land-use or ownership.

(b) Even if a standard codification list is available, it may be incomplete and not uniform throughout the State.

(c) Different Colloquial/local terms are used for indicating a characteristics of a plot (like soil/land-use, crops grown).

(d) Different practices followed over the decades in a locality are treated as a standard rather than the name list prepared by the States' Revenue Department.

(e) Different practices introduced by individual Patwari (village revenue functionary) are already recorded in the manual registers for the last few decades.

(f) Non-adoption of any scientific methodology to name a characteristics. For example the soil type of a plot is mentioned using different local names and it has no relation to soil taxonomy as adopted by different National level organizations like All India Soil and Land-use Survey (AISLUP).[6]

(g) In Land Records, data has to be preserved and the legacy data entered during the initial stages of computerization two decades back have to be migrated to the latest State of art technology without loss of the original data captured in the computer for the first time.

At the national level, DoLR constituted a committee of principal secretaries of some major States/UTs which requested NIC to undertake a study with the help of State NIC—Land Records teams to devise standard codifications for following fifteen attributes/parameters at the national level to make them interoperable with other domains such as Agriculture, Banking, Water Resources, etc.

- Location
- Crop Code
- Season Code
- Soil Type Code
- Source of Irrigation
- Land-use
- Area Unit/Extent
- Ownership Code
- Size of holding
- Mutation type
- Tenancy type
- Encroacher type
- Caste/Tribe
- Gender
- Encoding Standard

Structure of the Present Coding Scheme

Basic land records data is available in different registers as prescribed by State Government at District/Tehsil level. The number of these basic registers containing land records data is varying from State to State. However, there are master registers which contains the details of land ownership and plot details. These master registers are often known as khata register and plot register. These plot register contains the land records details of all the plots.[7]

The plot details include the soil type (black/red soil/alluvial), trees grown, types of crop (wheat/rice/maize) grown, land-use (barren/cultivable). The ownership type of the plot is specified like Government, Private or institutional. The khata register or khatauni contains details of ownership like name of the owner, father's name, caste and location. The format of khatauni with different codes is shown below[7] (Fig. 5.1).

All these physical attribute pertaining to ownership and plot as available on the basic land registers or the entities which describe a piece of land has been codified during the electronic data capturing process. However, while designing these code no attempt has been made to get uniformity across the Districts or across the States. Hence, any query on the database across the States may not give the desired result.

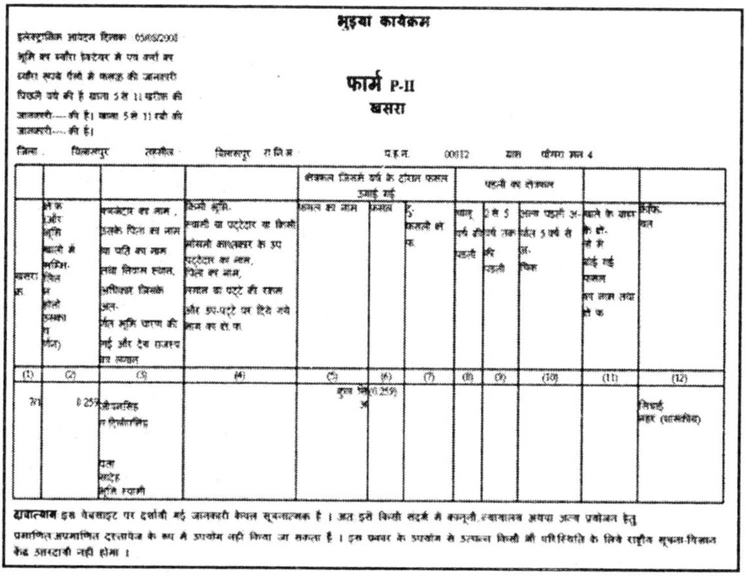
भुइयां कार्यक्रम

फार्म P-II

खसरा

Fig. 5.1: Details of a Khatauni (RoR) with Code Descriptions

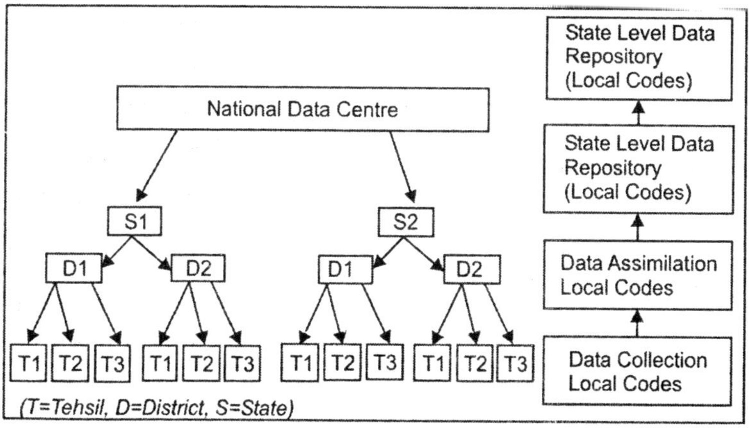

Fig. 5.2: Flow of Data from Tehsil to District, State and National Data Centre

Classification Methods of Coding Scheme

Coding schemes generally group together similar or related land attributes. The two common methods of classification are a simple exhaustive listing and a hierarchy.

- ***Exhaustive Listing***: This method uses a list of land attributes with sufficient codes to meet the application needs. This is the simpler of the two methods. This method is easier to customize for local conditions but is harder to share or aggregate. For example, typical urban and suburban jurisdictions may have up to 50 different codes for classifying residential uses and no more than one or two codes for all water-related uses.
- ***Hierarchy***: In a hierarchal coding system, all land related attributes are grouped into classes, and each broad land attribute category may contain additional subclasses or levels. In a hierarchal scheme there are 10 or 15 base classes for broad categories, such as Location, Crops, Season, Soil Type, Source of Irrigation, Land-use, Area Unit/Extent, Ownership, Size of holding, Mutation, Tenancy type, Encroacher type, Caste/Tribe, Gender and Encoding Standard.

Approach

At present each State is having computerized land records centres and these centres have been used for issuing records of right (RoR) to the public. During national level data collection ***no alteration will be made to the existing data*** and facility. Only the data as ported to the national level data centre will be dovetailed to have these new codes for the attributes as under consideration (Transformation approach). This will ensue the continuous availability of the existing facility to the public. The National level data repository will be only used for management information system and planning. All attributes in the National Level Codification will be described in English language only. In order to achieve this uniform coding scheme the following sequences may be followed:

(a) Study of the various Coding Schemes as available at District and Tehsil Level.
(b) Identification of attributes, which will be taken for Uniform Coding Scheme across the States.
(c) Preparation of Uniform Code List with code and description for all the attributes as selected in step 2.

(d) *Look up tables may be prepared for compatibility* with the national level code and the existing code.
(e) Extraction of meaningful information through land records databases as available in different States.
(f) Formation of a National Level Data repository with uniformity across the States.

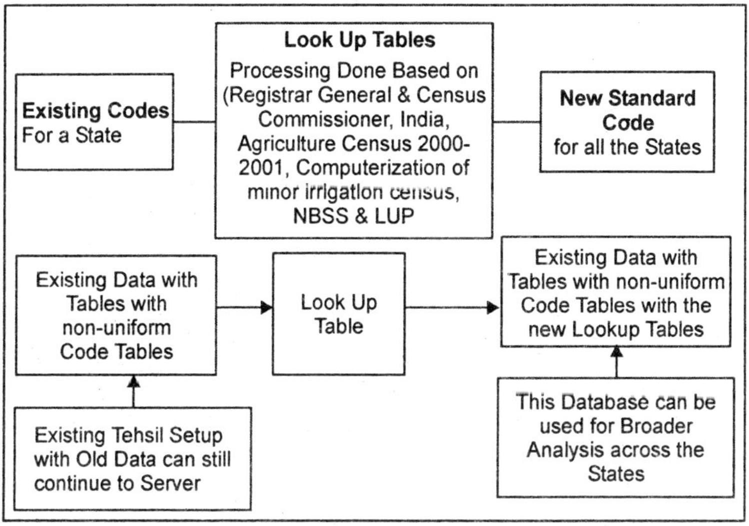

Fig. 5.3: Approach for Uniform Code[7]

Proposed Solution for Creation of Look Up Tables (Transformation Approach)

(a) Data available at Tehsil/Taluka Level where the project is operational.
(b) No physical connectivity exists at present between District and Tehsil level.
(c) District to Tehsil relationship is one to many. That is, a district can have more than one tehsil.
(d) The various categories of master codes or reference codes are classified into 15 categories such as crop, season, soil, source of irrigation, land-use, ownership, holding type, caste, encroacher type, gender, mutation, tenancy, in addition to location, area units and encoding.

(e) All Master Codes mentioned before are stored in a single table LRCodemaster tehsil-wise. The ID field (auto increment) is used as primary key which is referenced in transactional tables. Database constraints are enforced using the primary key-foreign key relationship.

(f) Variation of LRCodemaster from tehsil to tehsil within a district. Assuming District – 1 (D-1) has tehsils as Tehsil – 1 (T-1), Tehsil – 2 (T-2), Tehsil – 3 (T-3). T-1-LRCodemaster-T-1 (LRCT1), T-2-LRCodemaster-T-2 (LRCT2), T-3-LRCodemaster-T-3 (LRCT3). It has been observed that LRCT1, LRCT2 and LRCT3 are different in number of entries. Ideally LRCT1 = LRCT2 = LRCT3, but in practice, LRCT1, LRCT2 and LRCT3 are not similar. Hence need for LRCodemaster_std which is LRCStd at District level for the data merger utility to work.

(g) Building of LRCodemaster_std (LRCStd): After a detailed study of LRCT1, LRCT2...LRCTn, a LRCodemaster Standard (LRCStd) was prepared having 443 rows or entries for all the categories (excluding location for which the census 2011 was taken as the standard, area units having metric units as standard, data storage having unicode as standard) and approved. This was used by the data merger utility as reference. In any district D-1 having LRCT1, LRCT2, LRCT3, it was observed that in the worst case scenario, one of LRCT1 or LRCT2 or LRCT3 matched LRCStd in all aspects such as number of rows, entries, ID. Then that tehsil or taluk was chosen as reference database or destination database for that district in the State. The name of the database was *landrecords*. The remaining tehsil database was chosen as source database. The name of the source database was *landrecords1*. The data merger utility (DM) would compare LRCT1 = LRCStd of *landrecords* with LRCT2 of *landrecords1* and would merge (append) the data only in the following conditions: (i) If LRCT2 has extra rows in the end, that is if LRCT2 >LRCStd, (Assuming LRCT2 has 450 rows), or if LRCT2 = LRCT1 = LRCStd. Rest it would flash a message that "LRCodemaster of destination database and source database as different" and exit the programme.

(h) Prior to development of the utility, the master codes data of all the 35 States/UTs were analyzed for variation and also the pattern of storing the data too were studied. A need to eliminate human intervention was also suggested to make the utility automatic as far as possible.

(i) The Code checker utility developed was applied on the data collected from the States/UTs. The output reports were analyzed in depth. The variation was in the form of ID, Extra, Description, missing, blank.

   (i) *Extra*: The master codes needed to be appended to LRCodemasterStd of respective States in the local language.

   (ii) *Missing*: The master codes present in LRCodemasterStd but not being used in respective tehsils LRCodemaster (LRCT1) of the States/UTs.

   (iii) *Blank*: The master codes with description blank which can be deleted in LRCT1

   (iv) *Description*: Spelling mistakes, master codes could be dovetailed with existing codes of LRCodemasterStd like For example., Marushi Krushak can be dovetailed with Adhipatya Krushak in Owner type. Similarly Amrai can be dovetailed with Aam in Crop type, Mandir can be dovetailed with Dharmik Sthan in owner type. Hence human intervention is required.

   (v) *Empty*: Certain master codes found in LRCT1, but no entries are found for these codes in transaction tables.

   (vi) *ID*: ID field of LRCStd and LRCT1 has mismatch. This field is a Primary key and has a Foreign key relationship with transactional tables. This can be automated.

(j) Code Checker developed automates the steps mentioned above through a stored procedure (SP). This gives a list of codes in a particular category where human intervention is needed on the Left Hand Side (LHS) and the list of codes in LRCT1 on the Right Hand Side (RHS) and can be modified with the help of local revenue functionary. Code Checker helps to count the codes, reports for error, code correction of Extra, ID automatically, providing list of Extra Codes which can be appended/dovetailed with the help of local revenue functionaries.

Proposed Algorithm

For making LRCT1=LRCTStd after Code Checker reported LRCT1 not equal to LRCStd. Assume District D1 has 4 tehsils T1, T2, T3, T4 where n = 4. Code Checker—1 reported that LRCT1 = LRCTStd, but LRCT2 not equal to LRCTStd. Hence to make LRCT2 = LRCStd, the following steps were followed.

(i) The total number of IDs in LRCStd = 443
Margin for including Extra Codes
In all the 12 categories =156
Maximum number of IDs =599
The std IDs are as under 1-4→Caste, 5-8→Encroacher, 9-11→ Land-use, 12-14→Season, 15-16→Type of Holding, 40-44→Gender, 67-204→ Crop, 205-229→Source of irrigation, 230-238→Tenancy type, 239-328→Mutation type, 329-344→Owner type, 345-443→Soil.

(ii) Give temporary ID series for each of the 12 categories like
(1) Crop→600-699
(2) Land-use→700-799
(3) Caste→800-899
(4) Owner type→900-999
(5) Encroacher type→1000-1099
(6) Tenancy type→1100-1199
(7) Mutation type→1200-1299
(8) Type of holding→1300-1399
(9) Gender→1400-1499
(10) Source of irrigation→1500-1599
(11) Season→1600-1699
(12) Soil→1700-1799

(iii) Check for PK-FK relationships between LRCT1 and various tables. Disable relationships for inserts and updates of LRCT1 with tables through SQL Script.

(iv) Delete the blank entries in LRCT2 through SQL Script.

(v) Include Extra Codes of LRCT2 in LRCStd through SQL Script.

(vi) To overcome ID mismatch of LRCT1 with LRCStd, execute update SQL Script-1 for assigning temporary IDs and then execute update SQL Script-2 for assigning final IDs. These two SQL Scripts need to be executed for each category of ID mismatch as reported in all transaction tables.

(vii) For description with human intervention, try to dovetail codes as far as possible. If it is not possible to dovetail, include those master codes in LRCStd in the end.

(viii) Delete from LRCT2 through SQL Script. Import the data from LRCStd into LRCT2

(ix) Enforce relationship constraints again through SQL Script.

(x) Run Code Checker-1 again. It will give a blank report indicating that LRCT2 = LRCStd.

(xi) Repeat the steps for tehsils T3 and T4. Data merger utility runs smoothly once LRCT1 = LRCT2 = LRCT3 = LRCT4 = LRCStd

Procedure for Creating the Uniform Code for Location

Census 2001 database as the base. Various States have different hierarchy being followed. Some States have four levels and some States are having six levels. At the national level, we found that four levels such as State, District, Sub-District and Village which were common in all the States. The Sub-District level is known as Tehsil, Taluk, Revenue Circle or Mandal depending on the State.

Location master tables used in Land Records were collected from all the States. Some States had different tables for storing location code of district, sub-district, village, while some States stored them in a single table. The data collected from the States were mostly in database format (SQL Server) with English names in one column, but the rest in local language in 7/8 bit ISCII (Indian Standard Code for Information Interchange). The location master tables received from the States was treated as the source databases.

Stored Procedures were written to compare both the databases and generate scripts. Manual intervention was necessary for comparing the names of district or sub-district or village in both the source and destination databases. The estimated total number of revenue villages was 6.40 lakhs in the entire country when the land records databases were studied. The total number of villages according to Census 2001 database was around 5.9 lakhs. Revenue village directory is hosted on http://www.dolr.nic.in/dolr/land_reforms1.asp.

Conversion Utility for Conversion of Local Land Area Units to Metric System

A web based area conversion utility was developed which would convert the local land area units to metric system and *vice versa* upto five decimal places. It is hosted at http://www.dolr.nic.in/dolr/land_reforms1.asp in the heading Revenue Village directory and area conversion.

Utility for Migration of ISCII Land Records Data to Unicode

A small utility was developed for the migration of legacy land records data stored in ISCII 7 or 8 bit data into unicode. As the user interface is in local language and data is stored in ISCII 7 or 8 bit, States are slowly migrating to Unicode by using this utility.[8,9]

Challenges

(a) Large volume of data.
(b) Usage of local units in most of the States/UTs for area measurement instead of metric units.
(c) Lack of dedicated network connectivity between Tehsil to District.
(d) Local additions/deletions of master codes at Tehsils level were not discouraged.
(e) No proper database backup.
(f) Database may be corrupt in some locations.
(g) Low end systems were also used for processing large volume of land records data which is again time consuming.
(h) Lack of Annual maintenance contract of the Computer Systems and other related peripherals.
(i) Legacy land records data which are permanent records and need to be maintained and migrated with the advancement of technology.[10]

Conclusion

(a) Preparation of look up table for all the parameters at the national level.

(b) This will make the Land Records data interoperable with other domains such as Agriculture, Banking, Judiciary, etc.
(c) During this study, it was also found that reference standards existed for the land records attributes.
(d) Efforts are on to constitute Meta Data and Data Standards (MDDS) Committtee for Land Resources to ensure further critical examination of the proposed code and their standardisation through participatory approach with State governments and concerned central ministries including DIT.

References

1. *Land Records Computerisation, Technical Report and User Manual,* October 1996, Land Revenue Department, Government of Sikkim and NIC, New Delhi.
2. *Land Records Information System, Software Design Description*, June 1997, NIC, LRISD, New Delhi.
3. *Land Records Information System, Software Requirement Specification.* June 1997, NIC, LRISD, New Delhi.
4. *Computerisation of Land Records: A State-wise Profile,* June 1994, NIC, LRISD, New Delhi.
5. *Comparative Study Report: Land Records Computerisation,* e-GIFI, NIC, New Delhi, July 2004.
6. *NRIS Node Design and Standards, (NNRMS)*-ISRO, Bangalore, Februrary 2000.
7. *Study of Uniform Coding Scheme for Computerization of Land Records*, August 2008.
8. Various sources from the Internet.
9. *Computerisation of Land Records: National Perspective;* SD Meena *et al.* Land Reforms in India, Sage Publications, 2005.
10. *Social Impact of Computerisation of Land Records,* Vinay Thakur, D. Dutta, Ganesh Khadanga, D.S. Venkatesh. Adapting, e-Governance.

Chapter – 6

Leveraging Enterprise GIS for Land Record Management

Venkata Krishna R.

Introduction

Land continues to be a major source of government revenue through stamp duties, and is a key element in implementing a wide range of government programmes. Land policies and institutions will have a far-reaching impact on the country's ability to sustain high rates of growth, on the degree that economic growth benefits the poor, and on the level and spatial distribution of economic activity.

Land comes under the State subject and all along we have been hearing land tenure policy, legal and regulatory reforms, land-related conflicts and dispute resolution, land titling and registration. There is consensus that land administration is in urgent need of improvement. How to bring about such improvement most effectively, to improve land access and productivity of land-use in a sustainable fashion are important questions. The issue is not 'whether' but rather 'how' to do so most effectively. It is pertinent to use Enterprise GIS for land record management for effectively managing the land record.

Statistics shows registering property in various economies as per the data compiled by World Bank as on June 2011, see Table 6.1.

From the above figure we can draw various conclusions; However, the purpose is that we have a huge scope to ease the property registering process and bring down the duration time of registering the property with the most efficient procedures through a single window system.

Table 6.1: Statistics Regarding Registering of Property

Country	*Procedures (Nos)*	*Time (Days)*	*Cost (% of Property Value)*
Brazil	13	39	2.3
Russia	5	43	0.2
India	5	44	7.3
China	4	29	3.6
US	4	12	0.8
UK	6	29	4.7
Australia	5	5	5.0
UAE	1	2	2.0
New Zealand	2	2	0.1
France	8	59	6.1
Germany	5	40	5.2
Japan	6	14	5.7
Spain	5	13	7.1
Singapore	3	5	2.8
Canada	6	17	1.8
Bangladesh	8	245	6.6
Pakistan	6	50	7.7
Sri Lanka	8	83	5.1

The Department of Land Resources in the Government of India is implementing the National Land Records Modernisation Programme (NLRMP) involving survey/resurvey of land using modern technology, computerization of land records, digitisation of maps, computerization of registration and mutation system and integration of all these into a seamless system with the ultimate goal of ushering in the system of conclusive titles with title guarantee. This seamless system calls for Enterprise GIS system for integrating the spatial data and textual data at the revenue office and registration office.

The records about the transaction are maintained by both the revenue department and the stamps and registration department. The overlap increases transaction costs and creates the potential for fraud resulting from inconsistent records. While merging the two departments is unlikely, clarifying the legal situation of land records and ensuring back-office integration will be essential and sharing the central database and application

Approach

Enterprise GIS talks of seamless integration of data and application with various users across the various functions in the revenue and registration department through a scalable, reliable, flexible and robust architecture connecting the tehsil or taluk or block and sub-divisional levels and monitoring cell at the State level.

An approach towards central Data updating and maintenance of land records to reflect ground realities in sync with ownership changes, ensure genuine land transactions and implement overall development programmes for effective enforcement of land reforms. It will minimize land disputes and create social harmony in villages. Creating and maintaining an effective land administration system is a multifaceted process often managed by many disparate agencies and organizations. Duplicated efforts and data quality problems are common. Although separate GIS applications have been used to manage individual aspects of land records in the past, the full potential of GIS can best be realized when applied across the entire cadastre workflow.

Enterprise GIS, with a geodatabase, data models, and an array of applications, provides the framework for an integrated workflow for creating and updating a cadastre that can be easily shared both within and between organizations. State land records administration and agencies using GIS can streamline workflows and enforce cadastral procedures in the correct and legal way through efficient job management. This delivers cost efficiencies and ensures that quality is maintained consistently throughout the process.

Today, the buying and selling of property involves a transaction on the land records cadastre. This transaction is important to public and private interests such as land registry agencies, banks, and title insurance companies. As the foundation of a free market economy with leveraged capital, many countries are adopting land privatization for taxation to fund government services and economic development.

GIS, with integrated surveying and database management, along with map production and modelling support, is considered the essential information technology for land records solutions (Fig. 6.1).

Land administration agencies must manage their land information assets efficiently and effectively by moving from legacy "silo" workflows to an enterprise solution. There is a need for creating a robust GIS data model, like geodatabase at the heart of a cadastral enterprise transactional workflow.

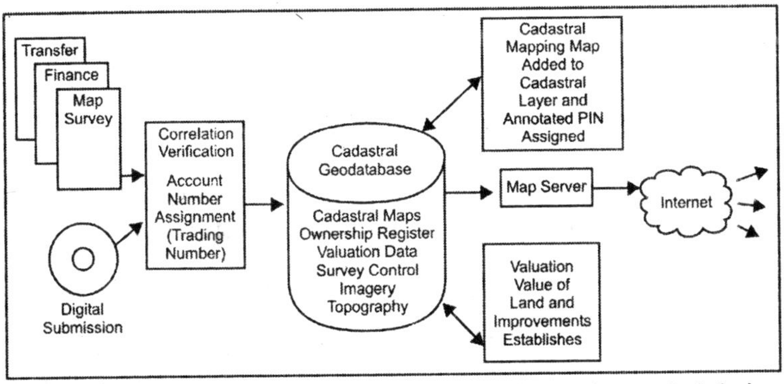

Fig. 6.1: Architecture of Modern Spatial Systems for Land Records Solution

The geodatabase with its associated ArcGIS software supports the land administration life cycle—data capture, data management and processing, and information dissemination—in an open environment that supports IT standards and interoperability with existing systems. An automated workflow modelled and implemented in GIS improves accuracy and quality and saves labour, time, and money. With the advent of band width and internet availability to different parts of the country, data can be delivered to public *via* websites. There is a huge increase in electronic transactions for the utility bill payment, etc., in urban areas, similar web based land transactions can substantially increase over the period; wherein over some time majority of Government interactions can happen through web. e-government will save taxpayers a significant amount of money and add value to their experience with government as their needs are better served.

GIS technology supports Web standards and is often the engine of government portals. The Internet has introduced a whole new way to interact with government. GIS makes that interaction more powerful and presents data in a form that makes viewing and understanding easy. State cadastre systems require many levels of technology that have the capacity to manage millions of land parcels and real estate objects. Esri's integrated collection of GIS products is well suited for these requirements. In addition, State cadastre systems are often decentralized and operate as an enterprise of multiple computers in many locations, requiring more than desktop applications. The Esri geodatabase template for land records supports the industry's efforts to standardize the way land records are defined and stored in data management systems for multiple users. This can be readily used to suite the various State requirements. Esri's ArcGIS for Server technology supports public access to land records and title registry *via* the Web.

Integrating GIS and computerized land titles enables the tax assessment function to be concurrent with spatial data that is relevant to the tax valuation model. It also supports the creation and maintenance of a more accurate land records basemap using the tools and functions of GIS and provides a single repository of parcel geometry and descriptive data supporting workflow, updates, and mass appraisal input. GIS adds value to computerized land title systems, such as an appraisal model, which can place added value on property that has, for example, frontage on a golf course or lake. If the parcel data and land-use data are maintained in a geodatabase, the frontage calculation is simple and easy using the spatial intelligence of the GIS.

Conclusion

Enterprise GIS way land record management shall meet the below objectives:

- Bring down the cost of delivery.
- Ensuring speed, accuracy, transparency and dispute resolution.
- Information empowerment of land owners.
- Providing fast and efficient retrieval of information for decision making.
- Achieving low cost and easily reproducible basic land record data for reliable and durable preservation.
- Value addition and modernization in Land Administration.

Notes

http://www.Worldbank.org/rural.
http://www.esri.com.

Chapter – 7

"HARSAC" How Geospatial Technology using for Land Record Reforms under NLRMP in Haryana

Sultan Singh

Introduction

Land, which is a scarce natural resource has been regarded as a measure of wealth, status and power from times immemorial. Any developmental activity is nearly impossible to conceive without taking land into consideration. Now, it is being widely regarded that the efficiency in land management is one of the indices of a nation's development status. It can also be said that the rights of the citizens to own private holdings and enjoy the wealth generated from the same constitute a very important facet of public administration. This is particularly true for India, which is predominantly an agricultural economy. As India is rapidly transforming into an industrialized economy and moving into Information age, it consequently suffers from maladies amongst which poor management of land records is a classic malady.

Land records in India, are said to have originated during the Mughal period, but were put on scientific foundation by the British. Akbar is considered to have initiated land surveys in India by *Raja Todar Mal* overhauled the revenue system of Akbar's Mughal empire. During the British time, large-scale cadastral surveys were conducted to determine the boundaries of villages through traverse method of surveying and extent of each individual landholding and to settle the crop-sustaining ability of different soils. This had been done to rationalize the levy and collection of land revenue from the landholders.

Following the recommendation of the 1904 Committee of Government of India, the cadastral surveys were delegated to the States. The States evolved their own legal system of revenue surveys. The Northern States of U.P., M.P., Bengal, Rajasthan, Punjab had 'graphical output' of surveys as the legal document whereas the South Western States recognized Field Measurement Book (FMB) as the legal document.

Later, in 1964, the first State Cadastral Surveys Conference was held in Mussoorie which was motivated by the concept that State Department of Settlement Surveys and Land Records should exchange ideas on the subject to provide technical solutions to, for example, town surveys, etc. The conferences have been regularly held but the progress has been rather slow in substance.

Government of India had taken up computerization of records and introduction of other relevant technologies in fairly enthusiastic manner around 1990. Government of India felt that Land Administration in India shall have to evolve procedures and methodologies consistent with the social dynamics of the day. Ministry of Rural Development (MRD) has been identified as the Line Ministry to implement the programme. Under the programme, States have successfully completed digitization of non-spatial land records of over 500 districts in the country under the programme called Computerisation of Land Records (CLR). Haryana has successfully captured the spirit of CLR programme and implemented HALRIS under CLR project all across the State.

Over the last few years Government of India has undertaken certain projects to be executed on a Mission mode programme under the National e-Governance action plan (NeGP) with the mandate to deliver identified services to citizens in an electronic form by the year 2007. Needless to say, computerization of land records has been taken up as one of the mission mode projects under NeGP. Ministry of Rural Development has been identified as a Line Ministry to co-ordinate the programme since CLR was already being implemented by MRD. As part of the process MRD constituted a Committee comprising of Revenue Secretaries from about 7 to 8 States to study and come up with recommendations for roll out of the programme in 2004. The Committee submitted a report to MRD in April 2005 with the recommendations that Integrated Land Information System be built for the country with an urgent need to digitize existing maps dealing with land records. It took a few years for Government of India to freeze the programme. Although, a few projects like Passport Services, Company Affairs, etc., under NeGP could achieve to deliver services land records computerization failed miserably largely because of the inability to conceive a vision to implement the programme

It is natural that a project with 450 million land parcels effecting the entire population will generate intensive debates on the approaches to adopt. However, the following are some of the salient points which have emerged from the various deliberations:

(a) Add spatial value to CLR programme;
(b) Existing spatial records be scanned (For Preservation);
(c) Existing spatial records be digitized;
(d) Update them for current details;
(e) Capture other details like cropping, land-use, soil cover, irrigation, etc.
(f) Integrate the various processes dealing with Land Administration and build an integrated land information system; and
(g) Deliver data to Citizens and other Organizations through all possible media.

It was only in August 2008 that Indian Cabinet cleared a programme for the country under the name and style called National Land Records Modernization Programme (NLRMP). The programme offered 3 approaches, viz., Aerial surveys, ETS surveys and hybrid approach using high resolution satellite imageries to States to implement the project. The final guidelines for implementing the programme were also issued last year and funds were also disbursed to States.

In the rapidly evolving global market, organizations are evolving systems to act quickly and efficiently. Organizations are seeking new ways to collect, manage and analyze data to retrieve meaningful information out of it. According to survey reports, hardly 3 per cent of available data is analyzed. It is estimated that 90 per cent of all government data has some kind of location component such as an address or a telephone or a PIN/ZIP code. Retrieval of geo-referenced information reveals spatial patterns and spatial linkages not seen on conventional statistical data. An increasingly large number of government organizations and business houses are trying to make use of this technology to derive meaningful information from the data.

Technology that facilitates this has been called as GIS. GIS—the acronym for geographic information system. It carries a lot of baggage. For most of its existence, GIS has been defined by a need to analyze map-specific data and constrained by the demands of complex data structures. Moreover, many people define GIS simply and narrowly as an automated map management.

However, Geospatial technology is not just about geography or maps; It is about visualizing and understanding relationships and information just below the layer of numeric data. For that reason, GIS has thrived for many years in several countries in key industries, including oil and gas, real estate, government, telecom, and utilities. More recent opportunities include diverse areas considering the coming together of many allied technologies like GPS, Networks, etc.

The growth of the GIS industry depended on how well it can move beyond these stovepipes and expand its role in decision support. GIS had to reinvent itself; a new role was required that exploits the technology's virtues and demonstrates its value to the decision-support process. In essence, GIS had to outgrow the purposes for which it was invented and take its place as a spatial decision support system, adding space and spatial data for critical analytic perspective for which no substitute exists. Land which is the foundation on which all other layers get added on assumed new significance. Considering the fact that India has over 450 million land parcels virtually affecting the entire population of over 1 billion people, developing a robust and transparent land records system required to adopt technologies bordering on the cutting edges. It has no other choice but to leverage the benefits of geospatial technologies to get rid of the land records system of the secrecy and lack of easy access and push the data into public domain. The challenge is to formulate a message that clearly communicates the value that these technologies bring to land administration and management. It must demonstrate to the IT community that a major part of the core technology necessary to integrate space is already in place: the RDBMS and Spatial Technology must seamlessly integrate with it.

GIS provides an integrated system for creating and updating the land parcels that can be easily shared both within and between organizations. A host of software work together to handle the entire work flow from data creation to information distribution in an environment that supports information technology (IT) standards and interoperability with existing systems. Enterprise GIS is, therefore, revolutionizing land records management.

How Geospatial Technology using for Land Record Reforms under NLRMP in State of Haryana

Haryana State that has about 7,085 Revenue village Estates covering an area of 44212 sq. km area having a population of 2,53,53,081 has about minimum 1,10,53,000 land parcels. These land parcels are depicted in about 62,000 Musavis and has about 20,850 sahedas locations. These Musavis were generated through a survey involving consolidation during the 1957-1961. The Jamabndi details were started to be computerised since 2001 that went on till 2006.

Being a Nodal agency for Geo-spatial Technology HARSAC initiated adding spatial value to CLR programme in 2007. It started by attempting a pilot project on Kamal village in Sirsa district bearing *hadbust* No. 259. It was later extended to the Kanungo or Revenue circle Rori covering 11 surrounding villages. The project was extended to Dabawali with 69 villages and later the entire district with 327 villages having Musavies 5,600 was covered. During this entire excercise HARSAC froze the methodology to be adopted for the State for building an integrated land information system for the State. Later, they took up work in a different topographical area of Ambala district with 495 villages and 2,800 Mussavis. This was followed up by taking up the same work in Hisar district with 265 villages with 5,000 *musavis* and Jind Tehsil in Jind district.

Keeping in view of the above aspects, Financial Commissioner, Revenue, Haryana has been in the fore front in the country in implementing the project for the State. Haryana Space Applications

Centre (HARSAC) has been made a nodal agency to coordinate the programme and process for implementing the project.

The State of Haryana has an integrated revenue system unlike most of the States in the country. Unlike the physical segregation of Registration, Revenue and Survey Departments in other States, Haryana has a system where the Director, Land Records is also the Survey and Settlement Commissioner and the Inspector General of Stamps and Registration and the process percolates down. For instance, Tehsildar will also do the role of Sub-Registrar on a rotation basis. This ensures greater coordination on transactions and better follow up on the different processes.

Haryana has a fairly evolved land records system. During the last consolidation time, land has been divided into uniform grids of one acre each called Kilas and each 25 Kila grid is called a Murraba. The measurements were carried out using Chains and the unit of measurement was the Foot Pound System. Each chain was a 55 feet one with 10 Karams. Each Karam was of 5.5 feet in length and was made of either 7 links or 8 links. Karam was the lowest unit of measurement. This system of dividing land into uniform grids helped in eliminating any major errors in measurement of Kilas.

After the consolidation, maps were generated for the entire State and the data is being managed in hard copies. The textual data like Jamabandi has been digitized. Now, Haryana State would like to convert the existing graphic land records also into a digital form and establish a linkage with the textual data and using high resolution satellite data would like to build a modern land records information system with the following scope.

1. Scanning of Village Maps (Musavis);
2. Digitization of Village Maps (Musavis);
3. Updating of Village Maps (Musavis);
4. Linkage of digital RoR data;
5. Scanning and linkage of textual records;
6. Ground Control Points;
7. Georeference of Satellite and Village maps;
8. Interpretation of missing data from Satellite data;
9. Ground Truth Surveys; and
10. Training.

As part of the initiative HARSAC attempt a comprehensive Proof of Concept (PoC) Study Berheri village in District Yamunanagar have been taken up as PoC villages.

Proof of Concept Study Objective

The objective of the project is to define a process for implementing the project under NLRMP for the State. The Scope of Work for the PoC was comprehensive with following approach:

1. Scanning of Village Maps (Musavis);
2. Digitization of Village Maps (Musavis);
3. Link non-spatial data to spatial data;
4. Update Village Maps (Musavis);
5. Scanning and linkage of textual records;
6. Mosaic adjacent Villages;
7. Register village map with high resolution satellite data;
8. Ground Control Points and Geo Reference; and
9. Analysis of Data.

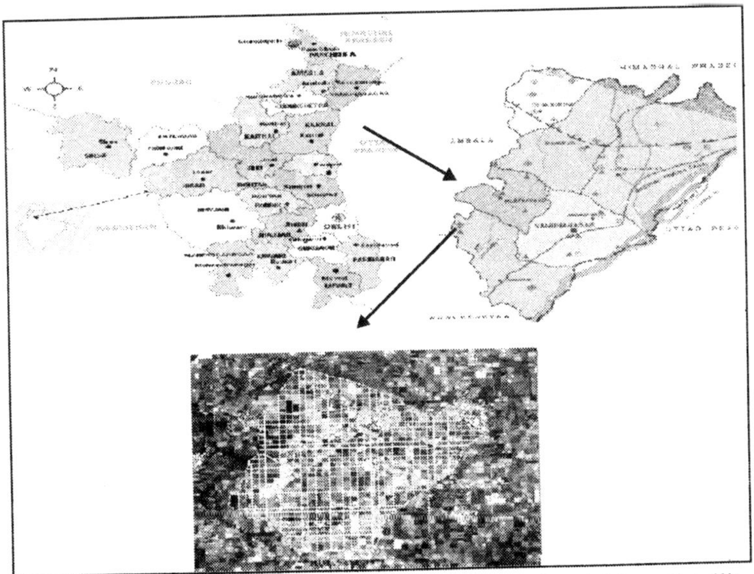

Fig. 7.1: Pilot Village Berheri Digitized Musavi Overlaid on Satellite Imagery

Apart from freezing the methodology to be adopted for the project, the secondary objective for the pilot village was to identify the Control Points and fix them.

Scanning of Village Maps

A village called Berheri was identified as the pilot village in District Yamunanagar. The maps pertaining to the pilot village along with the neighbouring villages were collected from the DC, Yamunagar office along with the textual records like Jamabandi, Misal Hakiyat, Field Books, etc. It was proposed to digitize the maps of neighbouring villages also since there was a problem with identifying the Control Points on the ground in this village. All the maps collected were scanned on either 200 dpi or 400 dpi depending on the quality of maps.

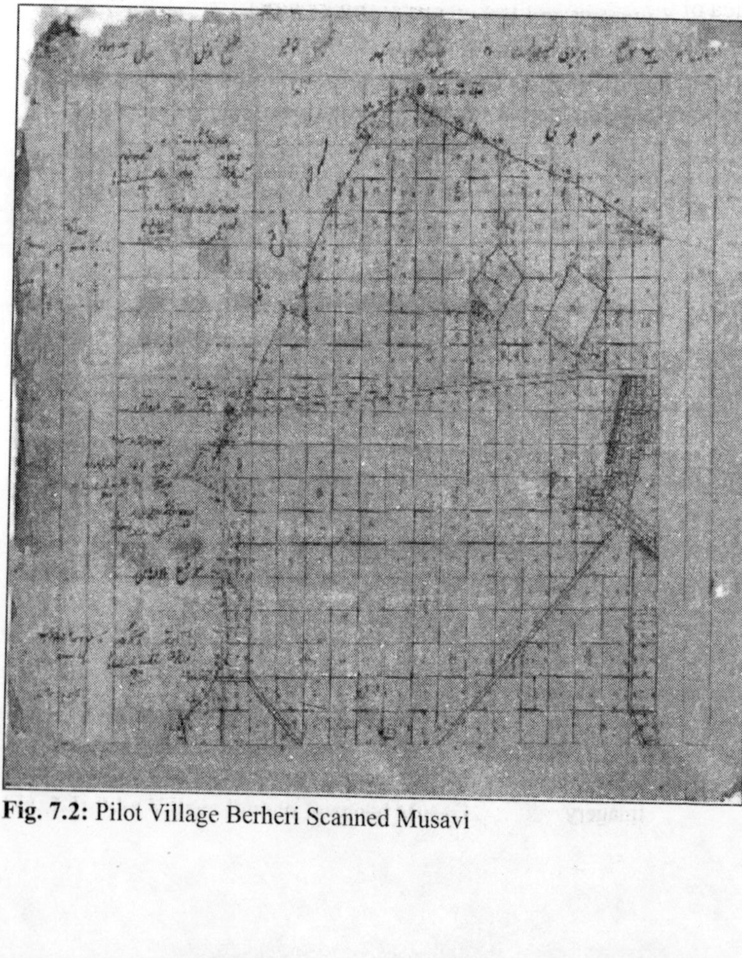

Fig. 7.2: Pilot Village Berheri Scanned Musavi

Digitization of Village Map

The process of digitization involved generating the Murrabas and Kilas using the dimensions for each Musavi (village map). Each Murraba is a rectangle of 180 Karams × 200 Karams. The scanned data was rectified using the Intersection Points of the Murraba Lines. 16 Points were taken to carry out the rectification of scanned Village Maps. Each Murraba consists of 25 Kilas. Each Kila is of 40 Karams × 36 Karams dimension. A grid was accordingly generated and all internal features and the village boundary were digitized using the dimensions mentioned in the Musavi. Wherever Cut Points or Arc Points were used in the Musavi, we had taken the dimensions and generated them. This will ensure that the digitized Musavi is an exact replica of the existing map. Line tracing was not carried out of any of the features on the map as it would lead to likely distortion in the dimension. Topology of the digitized features was built so that the attributes of the spatial features can be linked.

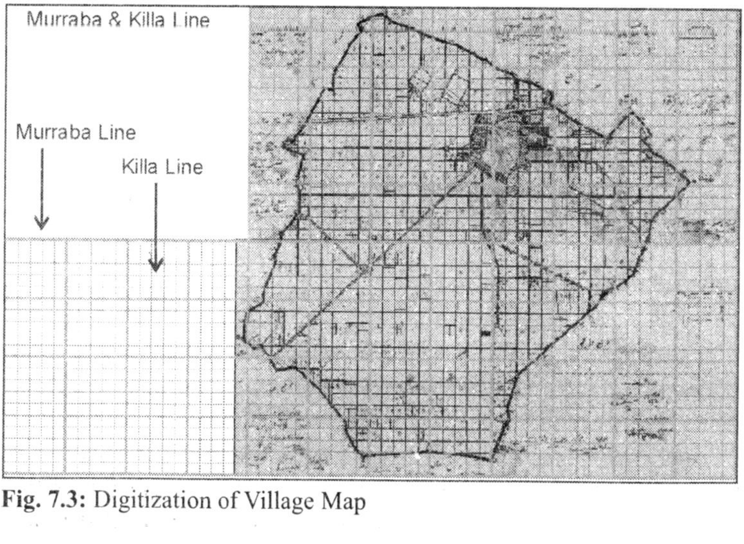

Fig. 7.3: Digitization of Village Map

The village maps of the pilot village with the neighbouring villages were generated in the years 1955-60 when Consolidation and Settlement for the village had taken place.

Linkage of Non-spatial Data to Spatial Data

We were furnished the digitized record of the RoR by DLR/NIC. The textual data was studied and after ascertaining the unique identifiers from the Tables we linked the textual data to the corresponding land parcels in the digitized village maps. Indian language support was provided in the GIS software so that the textual data that was in Devanagric script could be linked to the corresponding land parcels.

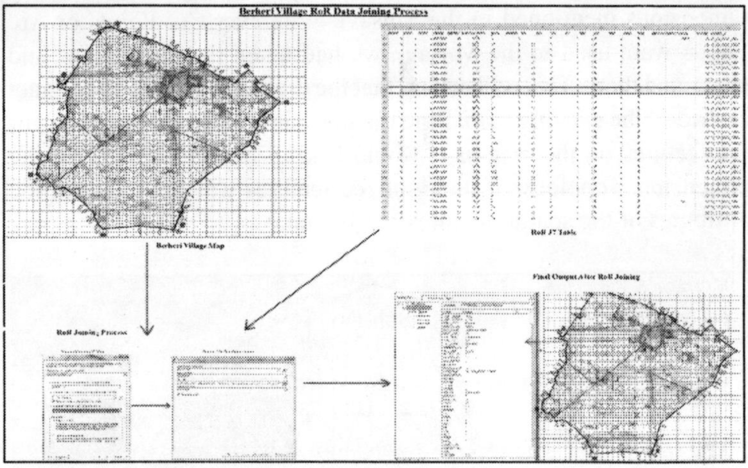

Fig. 7.4: Linkage of Textual Data to Spatial Data

Update Village Maps (Musavis)

The digitized village maps were printed and the printed copies were given to the Patwaris concerned for updating the changes that took place subsequent to the settlement process. These sub-division details that were available with the Patwaris were incorporated in the printed maps. These sub-division details were subsequently updated digitally. The RoR data was also linked again to ensure that the RoR data for the sub-divided parcels was linked.

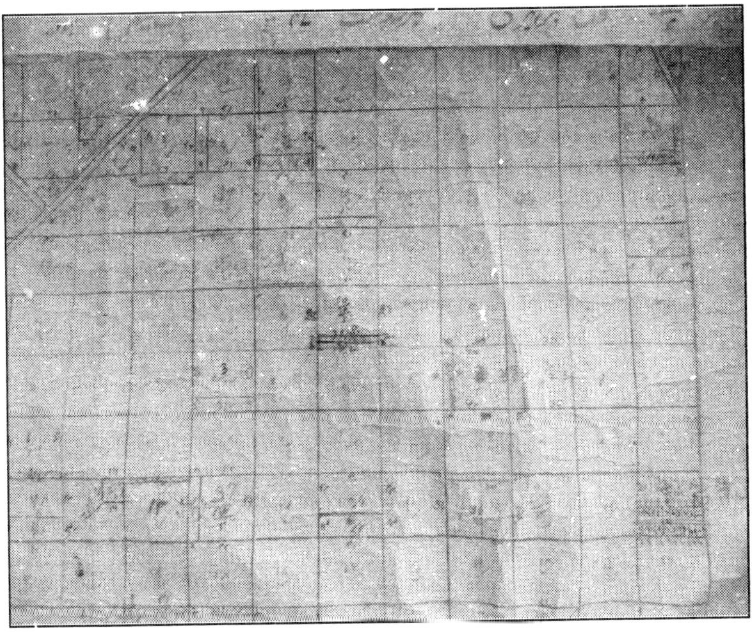

Fig. 7.5: Update of Musavis for Sub-Divisions

Scanning and Linkage of Textual Records

The textual documents like Jamabandi, Misal Hakiyat, Field Book, Sale Deed, etc., were collected from the Tehsildar offices in the district. These documents were scanned and linked to corresponding polygons. The following documents were scanned:

(a) Field Book—1954-55;
(b) Jamabandi—1962-63, 1968-69, 1973-74, 1977-78, 1983-84, 1988-89, 1998-99, 2003-04, 2008-09;
(c) Misal Hakiyat—1954-55;
(d) Intekal—1972-2011;
(e) Girdhawari—2004-08, 2009-2010; and
(f) Sale Deeds since 1998 to 2012.

Mosaic Adjacent Villages

The village maps surrounding the pilot village of Berheri were made

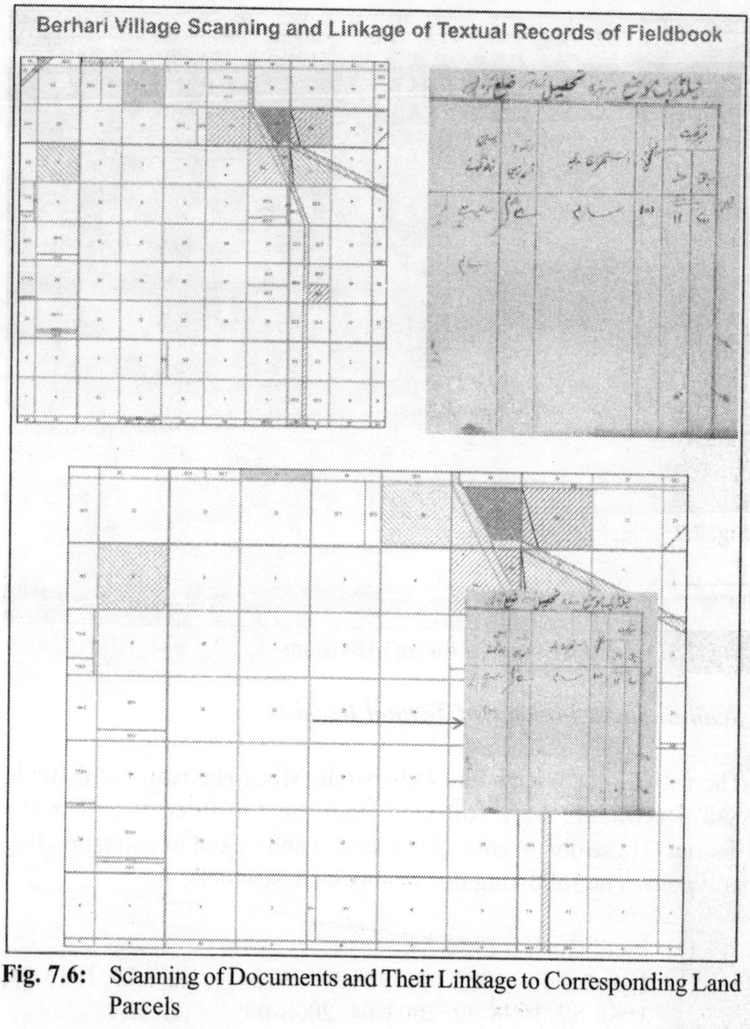

Fig. 7.6: Scanning of Documents and Their Linkage to Corresponding Land Parcels

into a mosaic. This was necessitated since some of the Saheda Stones in the field were missing according to the District administration. The mosaic of adjoining villages was carried out using two methods. One process involved matching the Murraba Lines and the other process involved matching the Burgi points. A report was also generated on the mismatch between the corresponding Burgis of neighbouring villages.

Fig. 7.7: Matching with Murraba Lines

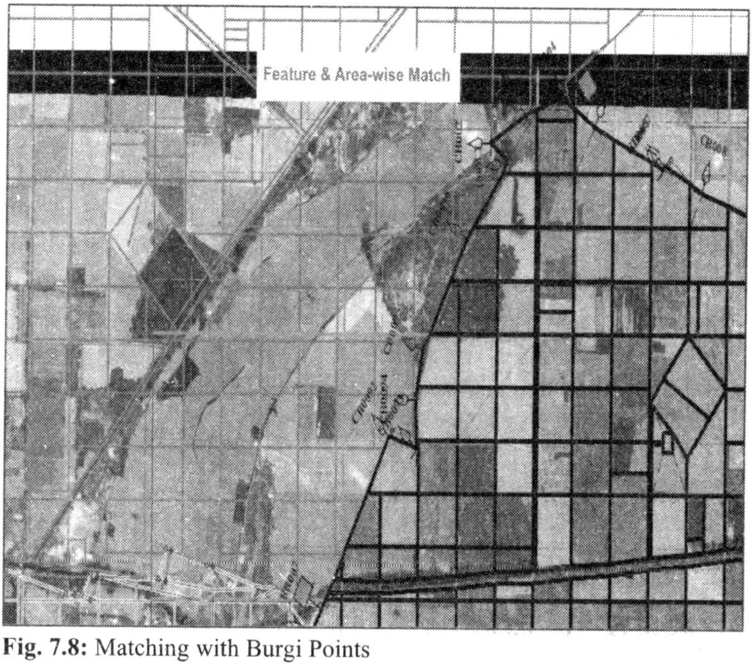

Fig. 7.8: Matching with Burgi Points

Register Satellite Data with Digitized Village Maps

We were furnished data from Cartosat and WorldView Satellites. Pan Stereo data of World View and Cartosat data were used and using certain identifiable reference features the satellite data was referenced with Village maps. Sahedas were marked on the Satellite data identifying them from the Village maps.

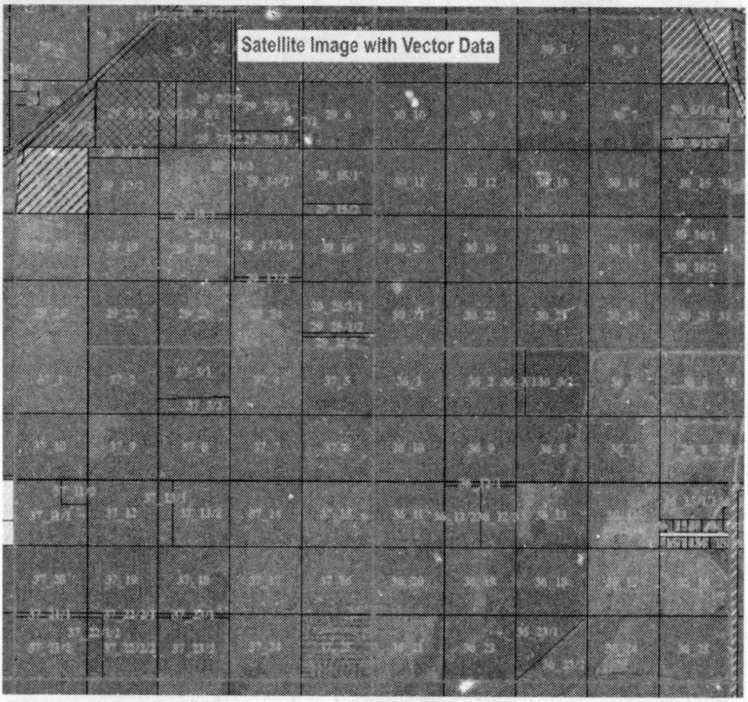

Fig. 7.9: Musavi Registered with Satellite Data

Ground Control Points and Geo Reference

It is proposed to carry prints of the village maps superimposed on the satellite data after demarcating the Sahedas, Burgis, etc., on the satellite data. Approximate location of Ground Control Points are proposed to be identified in advance. A field survey is proposed to be carried out to locate tertiary control points (TCPs) for further geo referencing of satellite data. The TCPs are proposed to be captured with reference to the PCPs and the SCPs. The proposed time of observation of TCP is 45 minutes using GPS in a differential mode.

Analysis of Data

A discrepancy report is generated between the recorded (RoR) area and the Computed area of each of the land parcels. The discrepancy is categorized into those with less than 1 marla (25.2928 sq. m is minimum area deviation accepted) variation and those above the limit of tolerance of 1 Marla. Those parcels that exceed 1 marla difference in area are further analyzed for possible reasons for the discrepancy in area. After geo reference using TCPs the exercise is proposed to be carried out again.

Sl. No.	*Accuracy Standards*		
	Dimensional accuracy		*Impact*
1	As Per Revenue System	No provision of the decimal system	Integer value of Karam is considered, i.e. smallest dimension is 1 Karam (1.6764 mts)
2	NLRMP Guidelines	Minimum 20 cm accuracy is desirable	± 8 inches is smallest dimension (0.20 mts)
Total Dimensional Accuracy Acceptable			1.8764 mts
	Area Accuracy		*Impact*
1	As Per Revenue System	One Marla is Acceptable Minimum area Due to Dimensional acceptable limit 1 Karam is acceptable	25.2928 sq.m is minimum area deviation accepted 2.810 sq.m minimum area deviation acceptable
2	NLRMP Guidelines	Due to 20 cm accuracy acceptable	0.04 Sq.m is minimum area deviation acceptable
Total Acceptable Area Deviation			28.1428 sq.m

The following analysis was carried out on the pilot study:

1. Mismatch between Map data and RoR data;
2. Mosaic of Village Maps;
3. Area validation of the village;

4. Burgi to Burgi difference in different Records;
5. Identification of Muraba Points;
6. Distortion in area in Arc Points; and
7. Public Property.

Mismatch between RoR and Map Data

- Total No. of Khasras in RoR: 1252
- In RoR:1366 (1252 if Min is taken into account).
- Khasras missing in RoR: 16
- Khasras missing in map: 15

Discrepancy Report on Computed Area of Musavi and RoR Area

S. No.	*Zero Tolerance*	*± 1 Marla tolerance*	*± 2.0 Marla tolerance*	*± 3 Marla tolerance*	*± 4 Marla tolerance*	*± 5 Marla tolerance*	*± 5 Marla tolerance*
1	1047	76	26	22	12	6	47

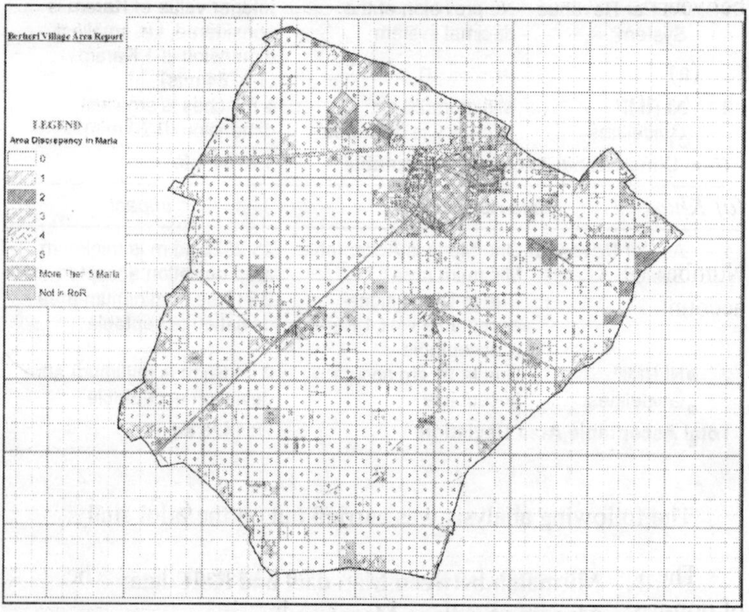

Fig. 7.10: Village with Land Parcels Highlighted for Discrepancy in Area

It is apparent from the above Table that about 9 per cent of the land parcels in the village have a discrepancy in area exceeding 1 Marla tolerance when compared between the graphic and textual records. The current system provides for a tolerance of 1 Marla in area computation to cover for the rounding off of Sersahis. It amounts to about 25 sq. m in area. The system also provides for accommodation of 1 Karam in linear measurements. These records in the map and RoR have to be checked by officers concerned.

The figure was higher at about 10 per cent but a QC of the digitization on the mismatch in parcels helped in removing the digitization errors. Mostly, the correction will be on taking the inner or outer arc point and that leads to change in area.

> *It is important that a Discrepancy Report be generated on the digitized data between the RoR data and the Musavis. The report will help identify possible digitization errors.*

It was observed that the following discrepancies were found between the Musavis and RoR data.

(a) Khasra Nos. found in RoR but missing on map,
(b) Khasra Nos. found in map but missing in RoR, and
(c) Duplicate records in RoR.

(a) Khasra Nos. found in RoR but Missing in Map

Number of Khasras in RoR: 1366 (However, if Min is taken into account, the Number of Khasras will come down to 1252).

- Matched in Musavi: 1236
- Not Matched: 15
- Khasra Nos. missing in Map:
  0-52, 14_8, 17_13, 22_17/1, 22_3, 22_8/2, 22_9/1, 23_2/1, 37_15/1, 37/15/2, 46_15, 46_6/1, 6_13/6, 6_26/2, 6_26/6

The possible reason for missing of these Khasras could be because of non-updating of map. DLR has to look into it.

(b) Khasra Nos. found in Map but missing in RoR

- No. of Polygons in Map: 1252
- Matched in RoR: 1236 (If Min is taken into account)
- Not Matched: 16
- Khasra Nos. missing in RoR:
  48_15, 48_6/1, 36_15/12, 38_15/2, 38_15/1, 32_12, 33_2/1, 23_19, 22_17/1/1, 17_13/1, 17_13/2, 14_8/7, 15_5/2, 6_24/2, 6_24/6, 6_23/6

It was observed that 16 Khasra Nos. available in map were missing in RoR. DLR has to look into it:

(i) Map updated for sub-division but not reflected in RoR (4 records – 14_8, 17_13, 22_17/1, 22_3),

(ii) Error in sub-division data entry (2 records – 22_8/2, 22_9/1),

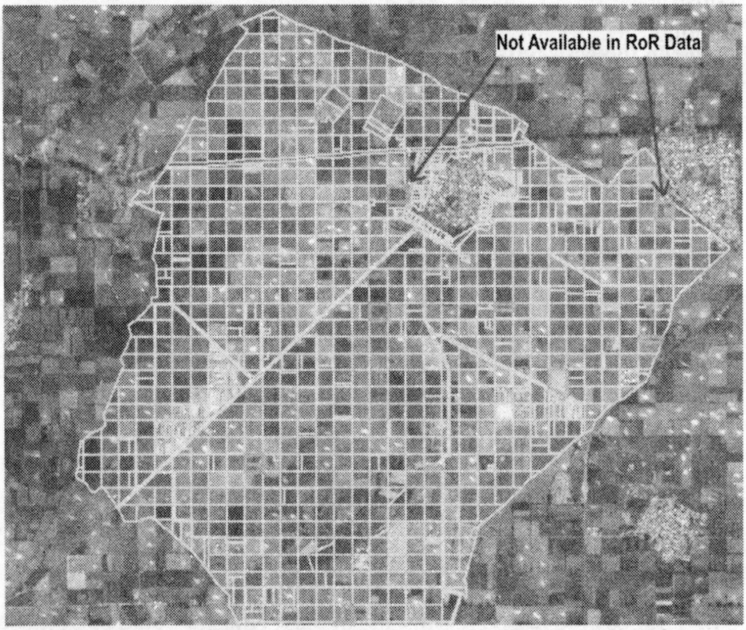

Fig. 7.11: Khasras Missing in RoR

(iii) Data Entry errors in Jamabandi (6 records – 23_2/1, 37_15/1, 37_15/2, 46_15, 46_6/1, 6_13/6),
(iv) Error in Misal Hakiyat (2 records – 6_26/2, 6_26/6),

It was observed that 15 Khasras were missing in the RoR that were available in the map. This could have happened because of non-updating of RoR.

(c) Duplicate Records in RoR

It was observed in the pilot village that about 224 RoR records are duplicate records. Some of the duplicate records have different areas recorded for the same Khasra Numbers and some have same areas. However, it was realized that if Min is taken into account all the duplicate records get reconciled.

Min is a temporary demarcation made in the RoR records whenever there is a change in land-use. For instance, if a factory is constructed on a property that area is demarcated as a sub-division although temporarily. It could be a systemic issue and needs to be studied further. It needs to be understood why Jamabandi gets updated if Min is a temporary phenomena. It should be recorded in a separate document if it is not a legal sub-division. DLR has to look into it. We are in the process of interaction with DLR and shall soon be able to get a clarity on it and resolve it accordingly.

Mosaic of Village Maps

It was observed that the village boundary data had marginal (±1 Karam) discrepancy in the corresponding boundary segments between villages. This may have occurred because of different survey parties carrying survey of adjoining villages and lack of coordination amongst them. Also, 1 Karam (5.5 feet) of survey chain was taken as a unit and part of it was not considered, i.e., decimals were not considered. This would have also led to the discrepancy amongst adjoining villages data. Because of the discrepancy it is inevitable that there will be marginal overlap or gap amongst the neighbouring villages when you make a mosaic.

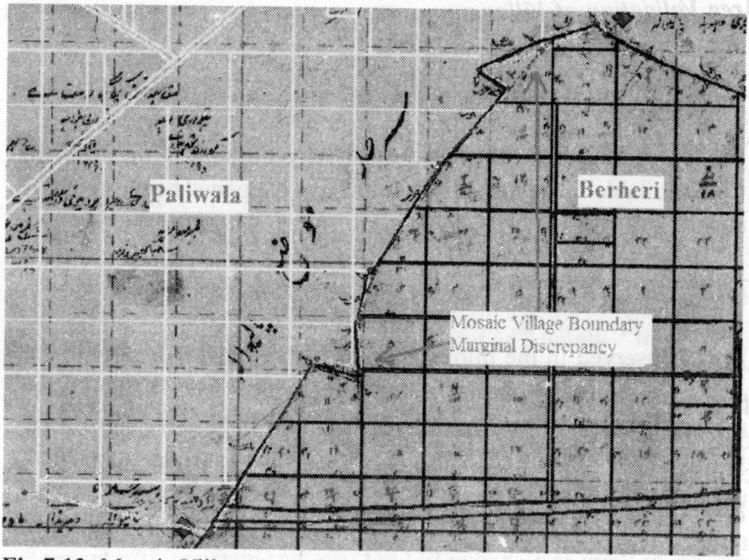

Fig 7.12: Mosaic Village Boundary Marginal Discrepancy

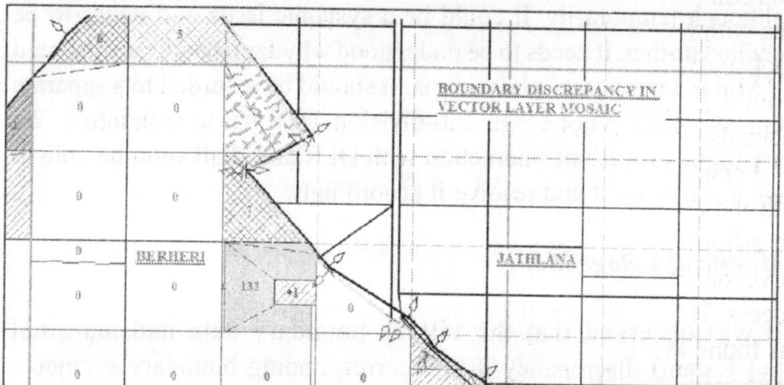

Fig. 7.1.3: Mosaic Village Boundary Minor Discrepancy

Area Validation of Village

Area from Different Records for Validation

S. No.		Musavi in acres	RoR in acres	Difference in acres	Jamabandi in acres	Field Book in acres	Misal Hakiyat
1	No. of Polygons	1252	1366	—	1313	1140	1140
2	Area	817.36	816.89	0.47	819	819	819

Area of Village: The total computed area of the village matched with the RoR area with about 0.47 acres difference over 817 acres. The Map area also matched with other records like Jamabandi, Field book, Misal Hakiqat with a difference of 1.64 acres. **This is a remarkable match considering the negligible variance. Apparently, the quality of survey was of a reasonably high standard.** Although the computed perimeter of the village was higher by over 21 metres compared to the recorded area on the Musavi, the total computed area of the village land parcels matched with the RoR area with hardly 0.5 acre difference. When compared with the Field Book and Misal Hakiyat the recorded area was found to be higher by about 1.5 acres. The mismatch is largely because of ignoring the decimals.

Perimeter of Village: The perimeter of the pilot village matched with the cumulative measurements of the corresponding Burgis of the adjoining villages with **remarkable accuracy**. Apparently, Surveyors coordinated well during the Consolidation and Settlement process. There were 7 villages surrounding the Berheri village and it was found that there was a mis match of 1 Karam with reference to the corresponding segment of only 1 village. This goes to prove that the surveyors coordinated well during the Traverse surveys.

However, there was over 21 m deviation with respect the computer perimeter. This could have occurred because of ignoring the decimal values. This deviation could reflect itself during the Mosaic process leading to transformation in areas at the time of geo reference of Musavis

The Sahedas were located on the satellite data by interpreting it and it was observed that there was a deviation of over 1 Karam of some of the Sahedas with reference to the ones on the Musavi. It is also **remarkable accuracy**

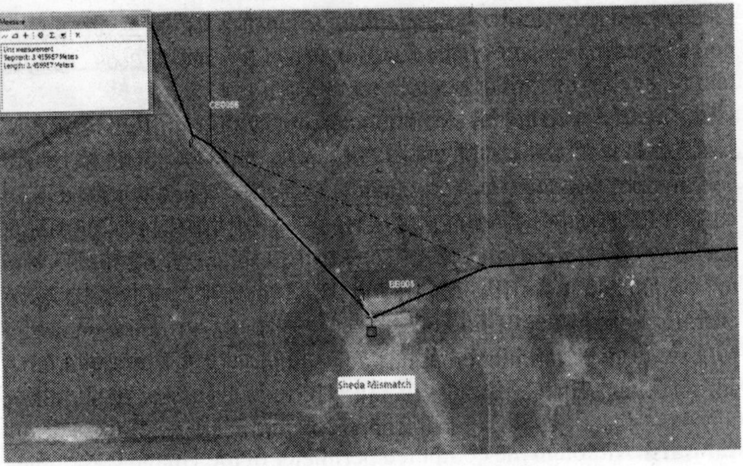

Fig. 7.14: Mismatch of Saheda

Burgi to Burgi Difference

We looked into the difference in the distance between the control points of the neighbouring villages. It was observed that there was a difference of maximum of 1 Karam in Burgis with reference to the distance in the corresponding Burgis of adjoining villages. This means that the Survey parties coordinated well amongst themselves to ensure that the Burgi matched well with adjoining villages and acted as a good control network. We also compared them using the field book.

We compared the total length of the perimeter of the village Berheri with the common boundary of the neighbouring villages and it was found that the difference in length was 1.67 metres over a length of 8013.33 m. We also took the computed distance of the perimeter of Berheri village and was around 8035 m. The difference between the computed distance of Vector map and the Burgi distance from Musavi was around 21.67 m which is about 0.27 per cent of the Berheri perimeter. This has happened largely because of the rounding of the decimal values.

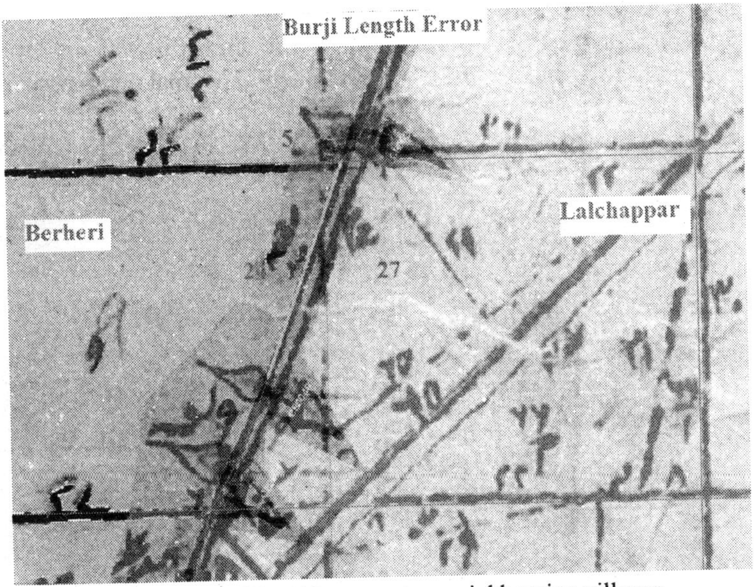

Fig. 7.15: Mismatch of Burgi between two neighbouring villages

It is very important that Quality Check be carried out of the digitized data for discrepancy between the Burgis. This will help as a very good quality check tool to detect and rectify digitization errors. Patwaris must promptly update Form 11C so that Vendors can incorporate the changes at the earliest.

Identification of Murabba Lines Intersection Points

During the consolidation time, Burgis were set up at the Murabba Lines Intersection Points also. It is possible that these Burgi monuments may have been destroyed for various reasons like floods, agricultural practices, etc. To determine the location of these points, Musavis were superimposed on satellite data and the Murabba Lines intersection points were cross checked with the ground features from satellite data. These points can be taken as TCPs and their location values captured using DGPS. In the pilot it was observed that such proposed Burgis interpreted using satellite data have a remarkable fit with the ground features.

Specifically, in the pilot village there was a problem with reference to measurement of a property in the Abadi area. Government officers wanted to determine these Burgis to help them make measurement.

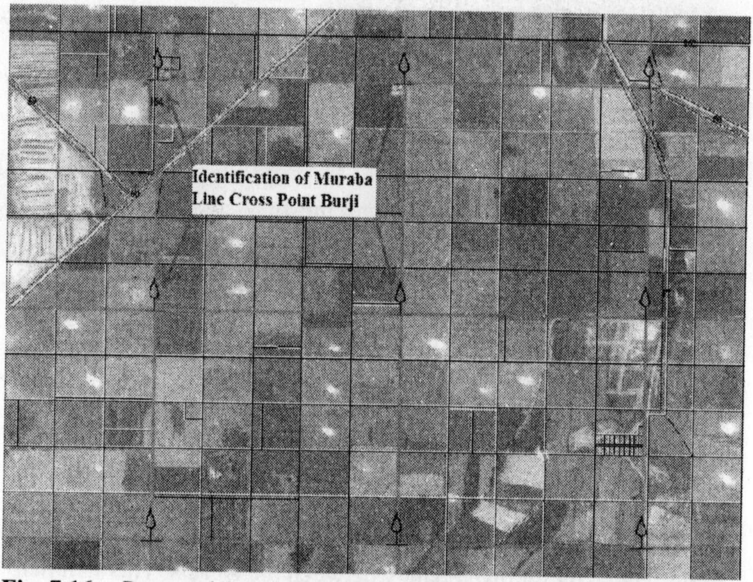

Fig. 7.16: Proposed Burgis at Murraba Intersection Points. The proposed Burgis fit very well with the ground situation

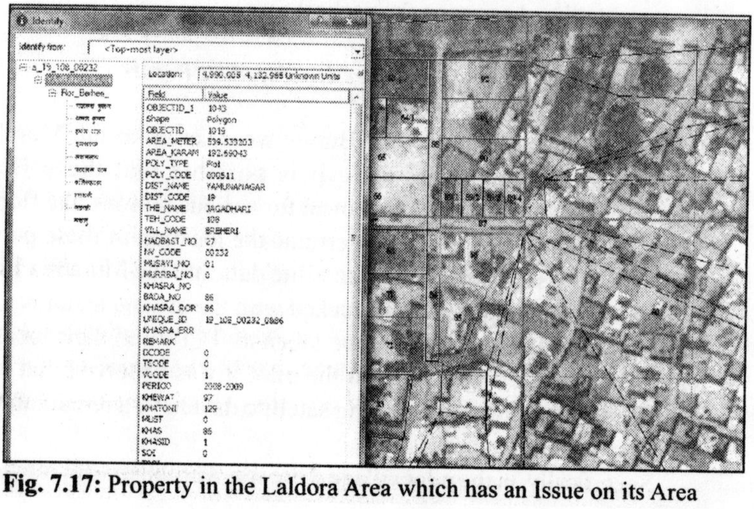

Fig. 7.17: Property in the Laldora Area which has an Issue on its Area

Distortion in Area at Arc Points

Generally, along the Village Boundary and around *Abadi* area Arc Points were used to fix points. Wherever, Points were generated using Arcs in certain situations a variation in area was observed as Arc Points which cannot be generated in such cases. Since decimals were ignored and because of the rounding of values arcs cannot be generated and to cut the arc decimal values have to be entered. For instance, if the Musavi depicts the two arc lengths as 12 and 13 Karams, it is possible that the actual distance on ground could have been 12.3 and 12.6. Since the distance is rounded off to the nearest Karam, it was depicted as 12 and 13. However, in a computer it is possible that the arcs may not intersect. In such cases, the GIS operator will increase the arc decimals proportionately on both the arcs till the arc intersects. It is possible that in certain cases in such areas there could be an area mismatch between the RoR area and the computed area.

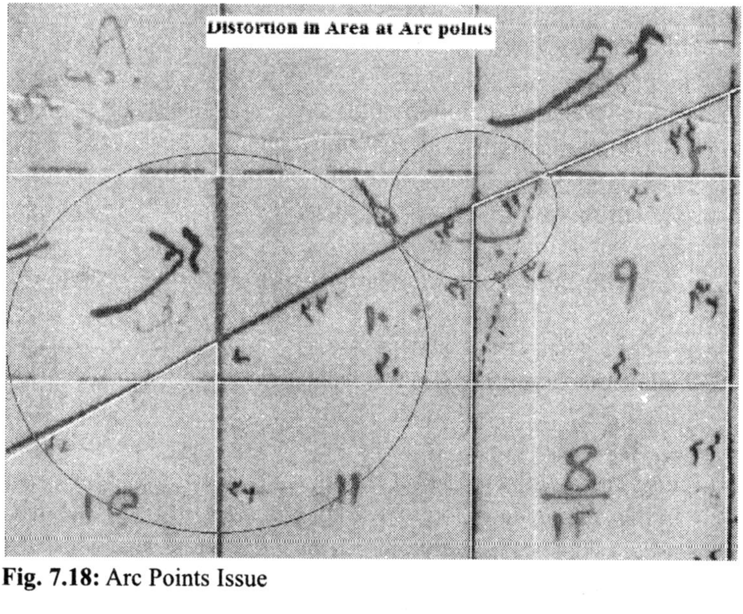

Fig. 7.18: Arc Points Issue

Public Property

It was observed that Public (Government) properties like roads and canals, etc., have a discrepancy in the RoR area and the computed area. This could have occurred because of the area computation process adopted at the time of survey. Area computation process involved breaking down the polygon into triangles and trapeziums and computing area mathematically. However, during this process Sersahis were rounded off. Each Marla has 9 Sersahis and anything above 4 was taken as 1 sersahi and below was taken as 0. This resulted in area mismatch in narrow land parcels as the rounding off has a huge impact because of the formation of multiple triangles and trapeziums. (See Table 7.2)

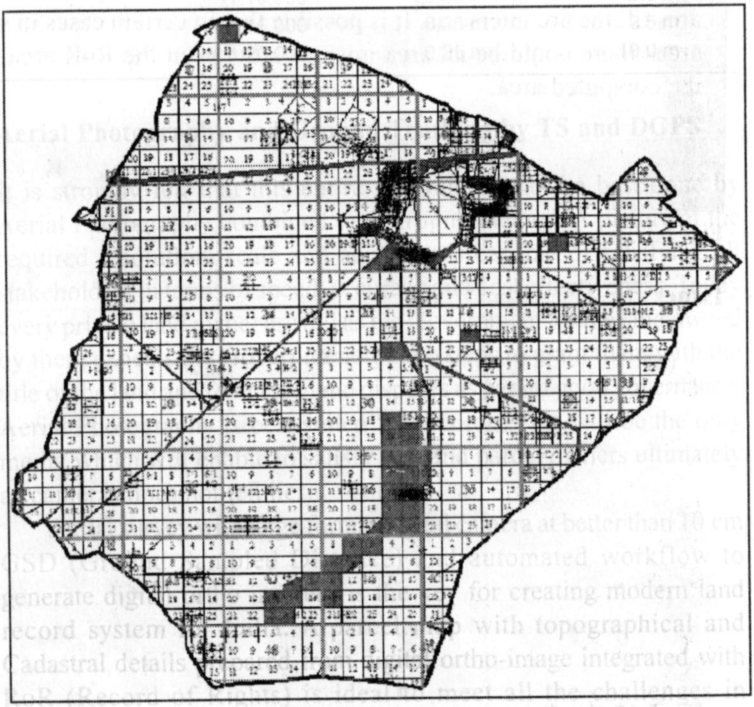

Fig. 7.19: Government Lands including Roads

Specifically, with reference to roads it was observed that almost all the roads have a discrepancy between the recorded area and computed area. (See Table 7.3)

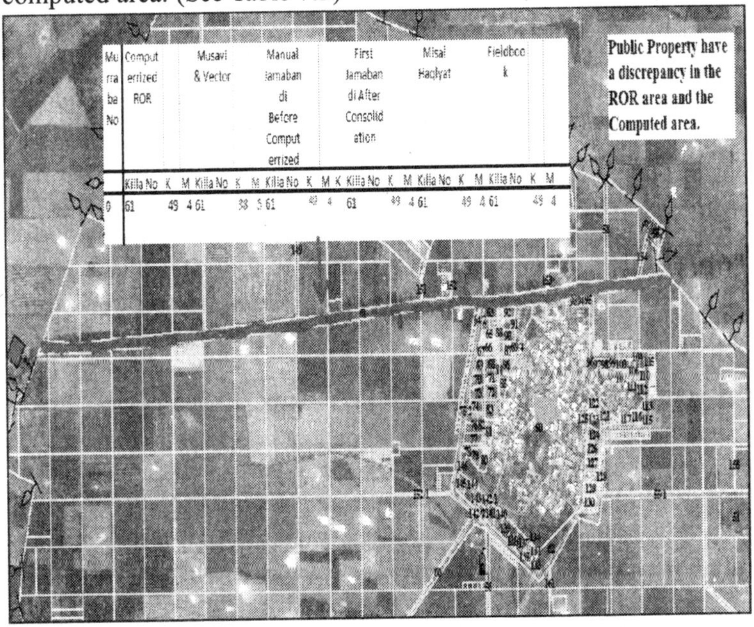

Fig. 7.20: Roads

Issues observed and understood from the Existing System Study

(a) **RoR Data:** Jamabandi Records pertaining to 16 Khasras were missing from RoR database. These records may have been missed because of some valid reasons. These need to be understood and action taken accordingly in future in the project. Also, the splitting of RoR data pertaining to a single Khasra into multiple ones needs to be understood and modified, if required (Min).

(b) **Missing Polygons in Map:** About 15 polygons were missing in the map. The RoR database reflects data for these land parcels. DLR has to look into it.

(c) **Area Discrepancy of Village:** There is bound to be marginal discrepancy in area of the village because of the following reasons.

(i) The distance between the Burgis is varying upto 1 Karam in a few of the adjoining Burgis. Although great care was taken at the time of survey to ensure match between the common Burgis of adjoining villages yet a few Burgis had difference. Apart from this the computed distance between Burgis and the recorded distance in Field Book or Musavi is varying because of ignoring part of a Karam. The field books were also checked in this regard.

(ii) It was observed that there was a problem with about 15 Khasras in the original records like village maps itself.

(iii) Public properties like roads and canals had a discrepancy. This could have occurred because of rounding off of Sersahis.

(d) **Discrepancy in Perimeter of Village:** It was noticed that the computed perimeter of the village exceeded by 21.79 m when compared with the recorded area on the map, i.e., the cumulative distance between the Burgis. However, the recorded distance matched with the distance from adjoining villages with only 1 Karam difference. But the issue of distortion of geometry of vector map because of over 21 m will have address during mosaic of maps of adjoining villages.

(e) **Discrepancy in Khasras along Boundary and Abadi:** It is quite likely that there will be marginal mismatch of Khasras falling along the village boundary and along the Abadi (Laldora). This is because of the rounding off of the distance with reference to the Arc lines.

(f) **Burgis (Control Points) are missing:** It was observed that the Murabba Burgis are missing in the pilot village and it is possible that they are missing in most of the villages in the Yamuna belt.

(g) **Discrepancies in different Records**: It was observed that there were discrepancies between different records wherever discrepancies were encountered. Although, the discrepancies were very small in number they need to be addressed and standardized. For instance, it was observed that there was variance in data between the Field Book, Musavi and Patwari map.

(h) **Change in Areas upon Geo Reference:** It was observed that there will be change in areas of Khasras upon geo reference of Musavis as the village maps undergo a transformation. It is for the DLR to take a decision whether satellite data is to be geo referenced or the Musavis are also to be done.

(i) **Mosaic of Sajra Maps:** Mosaic of village maps into Tehsil maps will not be possible without distortion. It was observed that there was variation in the distances amongst some of the corresponding Burgis and hence distortion is invariable when mosaic is carried out which means the areas of Khasras will also change. Also, the computed distance and the Musavi distance between Burgis varied because of ignoring the decimal value.

Recommendations

The above issues may give an impression that the existing data has issues. We would like to clarify that consolidation of land with a large unit of measurement of Karam ensured that the errors within a village are miniscule. Marginal errors are anticipated along the boundary because of ignoring part of a Karam. This is very negligible and the error can be accommodated by providing for a tolerance limit in area computation. This is necessitated because of rounding off of Karams and Sersahis. For example, the Revenue Acts of Andhra Pradesh, Gujarat provide for 5 per cent tolerance limit and an appropriate tolerance can be provided in Haryana too if the Revenue Acts do not provide for it already.

The current methodology being adopted by Haryana State suits it best keeping in view the fundamental aspect of consolidation and large unit of measurement of 5.5 ft. Almost 98 per cent of the land parcels in the village should not have much problems considering the large unit of measurement. However, it will have a problem as the very large unit of measurement will have a limitation because of the zig zag nature of the boundary. This problem can be overcome by simply ignoring the mosaic of village maps, Haryana State can implement the project without much hassles and must be continued aggressively.

Based on the pilot project analysis and the issues encountered the following approach for the project is proposed. All great projects or ideas are simple. This project also must be kept simple lest in the eagerness to implement a perfect system we end up adding too many features for the comfort of the project. From our association on some of the land records projects recently we have seen that in the eagerness to build a perfect integrated land records system in the country. A general tendency is to add any new initiative or with the sight of every new technology our approach to project changes. This temptation must be resisted. It must be borne in mind that our primary mandate is to push data into public domain and it must be the focus of this project. Keeping it in mind the following approach is being anticipated.

1. **Update RoR Data**. Certain Khasras depicted in the map are not available in the RoR database. In the pilot village Jamabandi Records pertaining to 16 Khasras were missing from RoR database. These records may have been missed because of some valid reasons. These need to be understood and action taken accordingly in future in the project. Also, the splitting of RoR data pertaining to a single Khasra into multiple ones needs to be understood and modified, if required (Min).
2. **Scan and Digitize Musavis**: Digitization of Musavis can be carried out as no major complications were encountered in the pilot nor are any huge challenges expected. However, Quality Check (QC) of digitized data is mandatory and adequate checks and balances shall be provided for quality check of data. The Department has to put appropriate mechanisms in place for QC of data.
3. **Update Musavis**: The digitized Musavis can be updated for subsequent changes. Patwaris can be issued prints of Musavis with RoR data linked to them. Initially, the focus must be on updating the Musavis and filling up of Annexure 11B for updating of Musavis process.
4. **Link Non-Spatial Data**: To the digitized Musavis, RoR data can be linked both prior to updating and after updating of Musavis. It was observed that duplicate records were

encountered in RoR and precaution shall be taken to ignore linkage of duplicate records with different areas.

5. **Scan Textual Records and Link Them:** Scan the textual records and link them to the corresponding digitized Khasras for documents retrieval system.
6. **Ground Control Points**: Collect the Tertiary Control Points (TCPs) and use them to geo reference stereo satellite data.
7. **Interpret for Missing Data**: Interpret the satellite data for missing portions (Khasras) of the map or other features like roads, Burgis, etc.
8. **Ground Truth Surveys**: The interpreted portions have to be validated through field surveys.
9. **Training**: Training has to be imparted on updating of Musavis and later on use of data.

Challenges and Limitations

(a) **Form 11B**: At the time of update of Musavis, Forms 11B is proposed to be filled by Patwaris. Form 11B will have the RoR area and Musavi area preprinted and data pertaining to Field Book, Misal Hakiyat, Jamabandi data after Consolidation and Jamabandi data before computerization needs to be filled by Patwaris for all the land parcels. Although, it is needed and important to be done, it could be a huge challenge for the Patwaris to devote so much time.

(b) **Mosaic of Village Maps**: Individual village maps presently treated as stand alone revenue estate entities. Mosaic of multiple village maps will be done as per the geometry and fit between adjoining villages is major task, because of the ignoring of the decimal values in a Karam (5.5 ft).

(c) **Measurements in Abadi Areas:** Measurements in Abadi areas could have error of 1 Karam using the current Burgis because of ignoring the decimal value. We mention this as we have observed that there is a 1 Karam deviation between the ground situation and the Musavi. This could have occurred because of ignoring the decimal value of Karam. The Department shall wait till the Control network under the current project gets generated and all measurements can be made using these Control points. The new Control data will help in achieving the accuracy desired for making measurements in Abadi areas. The new Control network shall help in achieving sub-meter level accuracy. Till then measurements in Abadi areas can continue as per the current system.

(d) **Geo-reference:** It is definite that there will be change in areas of all Khasras upon geo reference. Geo reference involves likely transformation of coordinates of Khasras. This may lead to change in areas even if it is very very marginal. The extent of transformation can be defined only upon completing the Control network. However, it is certain that the areas will change in excess of a Marlas or within a Marla in some parcels and it is a decision that the DLR has to take on declaration of new areas based on public perception.

Conclusion

It is apparent from the pilot project that considerable thought has gone into adopting the methodology for the project. DLR and HARSAC seem to have carried out extensive studies and pilot project to freeze on the methodology keeping in view the costs and time. It has been observed since the last 5 years much before the announcement of NLRMP that States had been struggling to capture the graphic data pertaining to land records all over the country. A few important factors seem to have been taken into cognizance while deciding on the approach for the NLRMP project for the State. Three of them seem very critical.

1. Firstly, land being made into uniform grids of 1 acre and an agglomerate of 25 of them being made into a Murabba seems significant. There seems to be hardly any issues in these land parcels as regards the dimensions. There is hardly any variance in the Musavi or on the ground in these areas. There is marginal variance on the boundaries and along public properties like roads and it can be accommodated by providing for a tolerance.
2. Secondly, there will not be a need to issue notices and call

for Objections and go through the adjudication process as the data being computerized is already settled. The computerization process can be relatively smooth sailing.

3. Thirdly, resurvey would involve capacity building. There is a shortage of qualified manpower either in the Government or in the private sector to attempt a resurvey project with the desired quality. It is possible to carry out a resurvey but to execute it with the desired quality to stand legal scrutiny is debatable. Capacities need to be built both in the Government and Private sector to carry out a fresh survey.
4. Resurvey would involve a long time frame. Invariably, fresh surveys will take not less than 7 to 8 years to map the State for land records in the current scenario. A project that involves such a long time span are difficult to sustain.
5. Even if a fresh survey is to be attempted, the current survey technologies require the existing data to be in a digital form. This project will lay the foundation for fresh surveys and will facilitate in carrying out fresh surveys with ease.
6. In keeping of Haryana study there is only 2 to 5 per cent area only require resurvey.

Keeping the above aspects in mind the project methodology is ideally suited for the State and must be persued aggressively.

Table 7.1: Duplicate Records in RoR

S. No.	MUST	KHAS	KNL	MRL	KNL & MRL	MURRABA_NO.	KHASRA_RoR	Vector_KNL & MRL	Vector Kanal	Vector Marla	Report	TOL	UNI
1	2	3	4	5	6	7	8	9	10	11	12	13	14
1	28	17	3	18	78	28	17	80	4		2	ì	28_17
2	28	17	0	2	2	28	17	80	4		78	µ	28_17
3	7	11	7	0	140	7	11	141	7	1	1	ì	7_11
4	7	11	0	6	6	7	11	141	7	1	135	µ	7_11
5	23	21	7	12	152	23	21	160	8		8	ì	23_21

(Contd.)

Table 7.1 (*Contd.*)

1	2	3	4	5	6	7	8	9	10	11	12	13	14
6	23	21	0	8	8	23	21	160	8	—	152	µ	23_21
7	29	15/1	6	11	131	29	15/1	133	6	13	2	ì	29_15/1
8	29	15/1	0	2	2	29	15/1	133	6	13	131	µ	29_15/1
9	29	5	7	15	155	29	5	160	8		5	ì	29_5
10	29	5	0	5	5	29	5	160	8		155	µ	29_5
11	30	1	7	12	152	30	1	160	8		8	ì	30_1
12	30	1	0	8	8	30	1	160	8		152	µ	30_1
13	30	10	7	12	152	30	10	160	8		8	ì	30_10
14	30	10	0	8	8	30	10	160	8		152	µ	30_10
15	36	19	7	11	151	36	19	160	8		9	ì	36_19
16	36	19	0	9	9	36	19	160	8		151	µ	36_19
17	36	20	7	11	151	36	20	160	8		9	ì	36_20
18	36	20	0	9	9	36	20	160	8		151	µ	36_20
19	36	21	0	9	9	36	21	160	8		151	µ	36_21
20	36	21	7	11	151	36	21	160	8		9	ì	36_21
21	36	22	7	11	151	36	22	160	8		9	ì	36_22
22	36	22	0	9	9	36	22	160	8		151	µ	36_22
23	43	1	7	18	158	43	1	160	8		2	ì	43_1
24	43	1	0	2	2	43	1	160	8		158	µ	43_1
25	43	6	7	11	151	43	6	160	8		9	ì	43_6
26	43	6	0	9	9	43	6	160	8		151	µ	43_6
27	36	12/2	3	14	74	36	12/2	75	3	15	1	ì	36_12/2
28	36	12/2	0	2	2	36	12/2	75	3	15	73	µ	36_12/2
29	15	14	9	12	192	15	14	193	9	13	1	ì	15_14
30	15	14	0	2	2	15	14	193	9	13	191	µ	15_14
31	29	11	7	16	156	29	11	158	7	18	2	ì	29_11
32	29	11	0	2	2	29	11	158	7	18	156	µ	29_11
33	29	26	0	9	9	29	26	18	0	18	9	ì	29_26
34	29	26	0	10	10	29	26	18	0	18	8	µ	29_26
35	25	25	7	6	146	25	25	148	7	8	2	ì	25_25
36	25	25	0	2	2	25	25	148	7	8	146	µ	25_25
37	28	3	3	18	78	28	3	80	4	—	2	ì	28_3
38	28	3	0	2	2	28	3	80	4	—	78	µ	28_3
39	25	23	7	12	152	25	23	160	8	—	8	ì	25_23
40	25	23	0	8	8	25	23	160	8	—	152	µ	25_23
41	25	13/2	3	8	68	25	13/2	136	6	16	68	ì	25_13/2
42	25	7	2	0	40	25	7	160	8	—	120	ì	25_7
43	25	13/2	3	8	68	25	13/2	136	6	16	68	ì	25_13/2
44	25	7	2	0	40	25	7	160	8	—	120	ì	25_7
45	25	14	7	16	156	25	14	160	8	—	4	ì	25_14
46	25	14	0	2	2	25	14	160	8	—	158	µ	25_14
47	25	7	4	0	80	25	7	160	8	—	80	ì	25_7
48	44	16	0	2	2	44	16	160	8	—	158	ì	44_16
49	44	16	7	18	158	44	16	160	8	—	2	ì	44_16
50	44	15	7	5	145	44	15	160	8	—	15	ì	44_15

(*Contd.*)

Table 7.1 *(Contd.)*

1	2	3	4	5	6	7	8	9	10	11	12	13	14
51	44	15	0	15	15	44	15	160	8	—	145	μ	44_15
52	43	19	4	0	80	43	19	160	8	—	80	ì	43_19
53	43	19	4	0	80	43	19	160	8	—	80	ì	43_19
54	29	8/1	6	2	122	29	8/1	124	6	4	2	ì	29_8/1
55	29	8/1	0	2	2	29	8/1	124	6	4	122	μ	29_8/1
56	20	13	4	0	80	20	13	159	7	19	79	ì	20_13
57	20	14	4	0	80	20	14	54	2	14	–26	ì	20_14
58	20	3	7	15	155	20	3	160	8	—	5	ì	20_3
59	20	3	0	5	5	20	3	160	8	—	155	μ	20_3
60	20	12	7	10	150	20	12	151	7	11	1	ì	20_12
61	20	12	0	2	2	20	12	151	7	11	149	μ	20_12
62	20	13	4	0	80	20	13	159	7	19	79	ì	20_13
63	20	14	0	4	4	20	14	54	2	14	50	ì	20_14
64	8	1	7	0	140	8	1	148	7	8	8	ì	8_1
65	8	1	0	8	8	8	1	148	7	8	140	μ	8_1
66	8	2	7	15	155	8	2	160	8	—	5	ì	8_2
67	8	2	0	5	5	8	2	160	8	—	155	μ	8_2
68	8	26	0	12	12	8	26	14	0	14	2	ì	8_26
69	8	26	0	2	2	8	26	14	0	14	12	μ	8_26
70	9	4	7	18	158	9	4	160	8	—	2	ì	9_4
71	9	4	0	2	2	9	4	160	8	—	158	μ	9_4
72	7	24/1	4	10	90	7	24/1	95	4	15	5	ì	7_24/1
73	7	24/1	0	6	6	7	24/1	95	4	15	89	μ	7_24/1
74	2	8	0	17	17	2	8	73	3	13	56	ì	2_8
75	2	8	1	16	36	2	8	73	3	13	37	ì	2_8
76	2	19	7	15	155	2	19	160	8	—	5	ì	2_19
77	2	19	0	5	5	2	19	160	8	—	155	μ	2_19
78	14	18	4	18	98	14	18	201	10	1	103	ì	14_18
79	14	21	4	0	80	14	21	160	8	—	80	ì	14_21
80	24	5	3	15	75	24	5	160	8	—	85	ì	24_5
81	24	5	0	5	5	24	5	160	8	—	155	μ	24_5
82	14	18	4	18	98	14	18	201	10	1	103	ì	14_18
83	14	20	4	0	80	14	20	160	8	—	80	ì	14_20
84	14	21	4	0	80	14	21	160	8	—	80	ì	14_21
85	24	5	4	0	80	24	5	160	8	—	80	ì	24_5
86	14	20	4	0	80	14	20	160	8	—	80	ì	14_20
87	24	15/1	3	10	70	24	15/1	79	3	19	9	ì	24_15/1
88	24	15/1	0	10	10	24	15/1	79	3	19	69	μ	24_15/1
89	24	4	7	10	150	24	4	160	8	—	10	ì	24_4
90	24	4	0	10	10	24	4	160	8	—	150	μ	24_4
91	24	7	7	10	150	24	7	160	8	—	10	ì	24_7
92	24	7	0	10	10	24	7	160	8	—	150	μ	24_7
93	24	13	7	15	155	24	13	160	8	—	5	ì	24_13
94	24	13	0	5	5	24	13	160	8	—	155	μ	24_13
95	25	15	7	15	155	25	15	160	8	—	5	ì	25_15

(Contd.)

Table 7.1 (*Contd.*)

1	2	3	4	5	6	7	8	9	10	11	12	13	14
96	25	15	0	5	5	25	15	160	8	—	155	µ	25_15
97	24	21	7	0	140	24	21	141	7	1	1	ì	24_21
98	24	21	0	1	1	24	21	141	7	1	140	µ	24_21
99	20	1/2	7	14	154	20	1/2	156	7	16	2	ì	20_1/2
100	20	1/2	0	2	2	20	1/2	156	7	16	154	µ	20_1/2
101	15	21/2	2	10	50	15	21/2	56	2	16	6	ì	15_21/2
102	15	21/2	0	10	10	15	21/2	56	2	16	46	µ	15_21/2
103	7	13	4	14	94	7	13	104	5	4	10	ì	7_13
104	7	13	0	10	10	7	13	104	5	4	94	µ	7_13
105	7	14	7	12	152	7	14	160	8	—	8	ì	7_14
106	7	14	0	8	8	7	14	160	8	—	152	µ	7_14
107	7	18	2	10	50	7	18	95	4	15	45	ì	7_18
108	7	18	2	1	41	7	18	95	4	15	54	µ	7_18
109	1	15	10	0	200	1	15	204	10	4	4	ì	1_15
110	1	15	0	4	4	1	15	204	10	4	200	µ	1_15
111	12	3	7	15	155	12	3	160	8	—	5	ì	12_3
112	12	3	0	5	5	12	3	160	8	—	155	µ	12_3
113	9	13	7	18	158	9	13	160	8	—	2	ì	9_13
114	9	13	0	2	2	9	13	160	8	—	158	µ	9_13
115	12	4	7	12	152	12	4	160	8	—	8	ì	12_4
116	12	4	0	8	8	12	4	160	8	—	152	µ	12_4
117	21	9	7	10	150	21	9	160	8	—	10	ì	21_9
118	21	9	0	10	10	21	9	160	8	—	150	µ	21_9
119	21	1	7	16	156	21	1	160	8	—	4	ì	21_1
120	21	1	0	4	4	21	1	160	8	—	156	µ	21_1
121	23	14	8	0	160	23	14	160	8	—	0	ì	23_14
122	23	11	7	12	152	23	11	158	7	18	6	ì	23_11
123	23	11	0	8	8	23	11	158	7	18	150	µ	23_11
124	22	1	7	0	140	22	1	160	8	—	20	ì	22_1
125	22	1	1	0	20	22	1	160	8	—	140	µ	22_1
126	22	19	7	15	155	22	19	159	7	19	4	ì	22_19
127	22	19	0	5	5	22	19	159	7	19	154	µ	22_19
128	22	22	3	11	71	22	22	151	7	11	80	ì	22_22
129	22	22	4	0	80	22	22	151	7	11	71	ì	22_22
130	6	13/2	2	19	59	6	13/2	62	3	2	3	ì	6_13/2
131	6	13/2	0	5	5	6	13/2	62	3	2	57	µ	6_13/2
132	23	20/2	3	12	72	23	20/2	80	4	—	8	ì	23_20/2
133	23	20/2	0	8	8	23	20/2	80	4	—	72	µ	23_20/2
134	24	16	10	10	210	24	16	220	11	—	10	ì	24_16
135	24	16	0	10	10	24	16	220	11	—	210	µ	24_16
136	36	3/1	0	10	10	36	3/1	20	1	—	10	µ	36_3/1
137	36	3/1	0	10	10	36	3/1	20	1	—	10	µ	36_3/1
138	36	8	7	11	151	36	8	160	8	—	9	ì	36_8
139	36	8	0	9	9	36	8	160	8	—	151	µ	36_8
140	23	14	8	0	160	23	14	160	8	—	0	ì	23_14

(*Contd.*)

Table 7.1 (*Contd.*)

1	2	3	4	5	6	7	8	9	10	11	12	13	14
141	22	18/2	4	18	98	22	18/2	102	5	2	4	ì	22_18/2
142	22	18/2	0	4	4	22	18/2	102	5	2	98	µ	22_18/2
143	2	10	7	10	150	2	10	159	7	19	9	ì	2_10
144	2	10	0	10	10	2	10	159	7	19	149	µ	2_10
145	14	24	11	9	229	14	24	229	11	9	0	µ	14_24
146	14	24	0	10	10	14	24	229	11	9	219	µ	14_24
147	25	1/1/1	0	13	13	25	1/1/1	53	2	13	40	ì	25_1/1/1
148	25	1/1/1	2	0	40	25	1/1/1	53	2	13	13	ì	25_1/1/1
149	48	13	8	14	174	48	13	175	8	15	1	ì	48_13
150	48	13	0	2	2	48	13	175	8	15	173	µ	48_13
151	48	9	7	15	155	48	9	160	8	—	5	ì	48_9
152	48	9	0	5	5	48	9	160	8	—	155	µ	48_9
153	27	25	7	18	158	27	25	160	8	—	2	ì	27_25
154	27	25	0	2	2	27	25	160	8	—	158	µ	27_25
155	32	16	7	0	140	32	16	152	7	12	12	ì	32_16
156	32	16	0	10	10	32	16	152	7	12	142	µ	32_16
157	8	15	10	7	207	8	15	217	10	17	10	ì	8_15
158	8	15	0	9	9	8	15	217	10	17	208	µ	8_15
159	40	6	1	11	31	40	6	33	1	13	2	ì	40_6
160	40	6	0	2	2	40	6	33	1	13	31	µ	40_6
161	12	12	7	18	158	12	12	160	8	—	2	ì	12_12
162	12	12	0	2	2	12	12	160	8	—	158	µ	12_12
163	19	1	1	0	20	19	1	62	3	2	42	ì	19_1
164	19	1	2	2	42	19	1	62	3	2	20	µ	19_1
165	39	6	7	15	155	39	6	160	8	—	5	ì	39_6
166	39	6	0	5	5	39	6	160	8	—	155	µ	39_6
167	39	7	7	18	158	39	7	160	8	—	2	ì	39_7
168	39	7	0	2	2	39	7	160	8	—	158	µ	39_7
169	42	8	7	5	145	42	8	146	7	6	1	ì	42_8
170	42	8	0	2	2	42	8	146	7	6	144	µ	42_8
171	37	15	7	18	158	37	15	160	8	—	2	ì	37_15
172	37	15	0	2	2	37	15	160	8	—	158	µ	37_15
173	37	18	7	18	158	37	18	160	8	—	2	ì	37_18
174	37	18	0	2	2	37	18	160	8	—	158	µ	37_18
175	37	19	7	18	158	37	19	160	8	—	2	ì	37_19
176	37	19	0	2	2	37	19	160	8	—	158	µ	37_19
177	39	3	7	17	157	39	3	160	8	—	3	ì	39_3
178	39	3	0	3	3	39	3	160	8	—	157	µ	39_3
179	27	4	7	18	158	27	4	160	8	—	2	ì	27_4
180	27	4	0	2	2	27	4	160	8	—	158	µ	27_4
181	39	2	6	10	130	39	2	148	7	8	18	ì	39_2
182	39	2	0	1	1	39	2	148	7	8	147	µ	39_2
183	28	16	10	15	215	28	16	220	11	—	5	ì	28_16
184	28	16	0	5	5	28	16	220	11	—	215	µ	28_16

(*Contd.*)

Table 7.1 (*Contd.*)

1	2	3	4	5	6	7	8	9	10	11	12	13	14
185	28	24	8	0	160	28	24	220	11	—	60	ì	28_24
186	28	24	3	0	60	28	24	220	11	—	160	ì	28_24
187	47	3	7	15	155	47	3	160	8	—	5	ì	47_3
188	47	3	0	5	5	47	3	160	8	—	155	µ	47_3
189	12	7	7	18	158	12	7	160	8	—	2	ì	12_7
190	12	7	0	2	2	12	7	160	8	—	158	µ	12_7
191	37	20	4	0	80	37	20	160	8	—	80	ì	37_20
192	37	20	3	18	78	37	20	160	8	—	82	ì	37_20
193	37	20	0	2	2	37	20	160	8	—	158	µ	37_20
194	6	8	5	0	100	6	8	108	5	8	8	ì	6_8
195	6	8	0	10	10	6	8	108	5	8	98	ì	6_8
196	6	19/1	1	0	20	6	19/1	70	3	1	50	ì	6_19/1
197	6	19/1	2	12	52	6	19/1	70	3	1	18	µ	6_19/1
198	2	24	7	15	155	2	24	160	8	—	5	ì	2_24
199	2	24	0	5	5	2	24	160	8	—	155	µ	2_24
200	6	17/2	2	10	50	6	17/2	51	2	11	1	ì	6_17/2
201	6	17/2	0	3	3	6	17/2	51	2	11	48	µ	6_17/2
202	37	11/1	6	18	138	37	11/1	139	6	19	1	ì	37_11/1
203	37	11/1	0	2	2	37	11/1	139	6	19	137	µ	37_11/1
204	35	13	7	0	140	35	13	160	8	—	20	ì	35_13
205	35	13	1	0	20	35	13	160	8	—	140	µ	35_13
206	31	21/2	4	0	80	31	21/2	88	4	8	8	ì	31_21/2
207	31	21/2	0	8	8	31	21/2	88	4	8	80	µ	31_21/2
208	44	19	7	18	158	44	19	160	8	—	2	ì	44_19
209	44	19	0	2	2	44	19	160	8	—	158	µ	44_19
210	31	8	0	3	3	31	8	108	5	8	105	µ	31_8
211	31	8	5	6	106	31	8	108	5	8	2	ì	31_8
212	29	3	2	15	55	29	3	220	11	—	165	ì	29_3
213	29	8/2	0	9	9	29	8/2	36	1	16	27	ì	29_8/2
214	29	9/1	1	17	37	29	9/1	143	7	3	106	ì	29_9/1
215	29	8/2	0	9	9	29	8/2	36	1	16	27	ì	29_8/2
216	29	9/1	1	17	37	29	9/1	143	7	3	106	ì	29_9/1
217	29	3	2	15	55	29	3	220	11	—	165	ì	29_3
218	29	3	2	15	55	29	3	220	11	—	165	ì	29_3
219	29	8/2	0	9	9	29	8/2	36	1	16	27	ì	29_8/2
220	29	9/1	1	17	37	29	9/1	143	7	3	106	ì	29_9/1
221	17	13	0	5	5	—	—	0	0	—	–5	µ	17_13
222	17	13	6	14	134	—	—	0	0	—	–134	ì	17_13
223	14	8	8	8	168	—	—	0	0	—	–168	ì	14_8
224	14	8	0	5	5	—	—	0	0	—	–5	µ	14_8

Table 7.2: Total Area of Public Properties RoR, Map and the Deviation

S. No	*Khasra_No RoR*	*RoR_Area_M*	*Vec_Area_M*	*Area_Diff*
1	165	15.00	154.67	139.67
2	23	147.00	146.67	–0.33
3	24/1	81.00	80.67	–0.33
4	166	7.00	7.33	0.33
5	18/2	108.00	108.00	0.00
6	18/1	52.00	52.00	0.00
7	17	160.00	160.00	0.00
8	54	80.00	80.00	0.00
9	53	80.00	80.00	0.00
10	14/2	84.00	84.44	0.44
11	15	160.00	160.00	0.00
12	11	160.00	160.00	0.00
13	10	160.00	160.00	0.00
14	9/2	122.00	118.44	–3.56
15	2	160.00	160.00	0.00
16	3	160.00	160.00	0.00
17	23/2	80.00	80.00	0.00
18	24	160.00	160.00	0.00
19	25	160.00	160.00	0.00
20	17	160.00	160.00	0.00
21	16	160.00	160.00	0.00
22	167	99.00	99.17	0.17
23	14	160.00	160.00	0.00
24	15/1/1	86.00	85.78	–0.22
25	168	48.00	48.67	0.67
26	7	160.00	160.00	0.00
27	6	160.00	160.00	0.00
28	10/2	56.00	56.00	0.00
29	4	160.00	160.00	0.00
30	5	160.00	160.00	0.00
31	1/1	56.00	56.00	0.00
32	24	160.00	160.00	0.00
33	25	160.00	160.00	0.00
34	16/2	71.00	71.11	0.11
35	17	160.00	160.00	0.00
36	163	37.00	36.67	–0.33
37	57	319.00	302.39	–16.61
38	164	36.00	36.33	0.33
39	24	160.00	196.00	36.00
40	162	30.00	29.78	–0.22
41	18/2	98.00	102.22	4.22
42	169	59.00	58.67	–0.33
43	16/3	20.00	20.00	0.00
44	59	150.00	128.21	–21.79
45	170	8.00	8.00	0.00

(Contd.)

Table 7.2 (*Contd.*)

S. No	*Khasra_No RoR*	*RoR_Area_M*	*Vec_Area_M*	*Area_Diff*
46	160	9.00	8.45	−0.55
47	159	18.00	17.78	−0.22
48	158	77.00	97.56	20.56
49	24/1	3.00	3.33	0.33
50	24/2	3.00	3.33	0.33
51	24/4	2.00	3.33	1.33
52	161	9.00	6.90	−2.10
53	132	4.00	4.68	0.68
54	131	5.00	5.45	0.45
55	133	3.00	5.17	2.17
56	156	35.00	15.33	−19.67
57	134	5.00	5.17	0.17
58	135	5.00	5.17	0.17
59	136	5.00	5.17	0.17
60	56	453.00	347.44	−105.56
61	24	229.00	229.85	0.85
62	60	318.00	363.51	45.51
63	51	142.00	142.22	0.22
64	55/1	134.00	127.87	−6.13
65	157	18.00	17.78	−0.22
66	13/4	7.00	6.67	−0.33
67	13/5	25.00	23.21	−1.79
68	13/2	42.00	42.00	0.00
69	13/3	14.00	14.33	0.33
70	152/1	23.00	23.52	0.52
71	78	9.00	8.83	−0.17
72	7/16	2.00	2.00	0.00
73	7/15	2.00	2.00	0.00
74	7/14	2.00	2.00	0.00
75	7/13	2.00	2.00	0.00
76	7/12	2.00	2.00	0.00
77	7/11	2.00	2.00	0.00
78	7/10	2.00	2.00	0.00
79	7/9	2.00	2.00	0.00
80	7/8	2.00	2.00	0.00
81	7/7	2.00	2.00	0.00
82	7/6	2.00	2.00	0.00
83	7/5	2.00	2.00	0.00
84	7/4	2.00	2.00	0.00
85	7/3	2.00	2.00	0.00
86	7/2	2.00	2.67	0.67
87	6/6	2.00	2.00	0.00
88	6/5	2.00	2.00	0.00
89	6/4	2.00	2.00	0.00
90	6/3	3.00	2.67	−0.33
91	118	5.00	3.98	−1.02
92	115	4.00	4.44	0.44

(*Contd*)

Table 7.2 (Contd.)

S. No	Khasra_No RoR	RoR_Area_M	Vec_Area_M	Area_Diff
93	119	4.00	4.04	0.04
94	117	5.00	4.98	–0.02
95	120	5.00	3.39	–1.61
96	114	4.00	4.44	0.44
97	116	5.00	5.50	0.50
98	82	11.00	11.98	0.98
99	113	5.00	5.22	0.22
100	5/19	2.00	2.01	0.01
101	5/18	2.00	2.00	0.00
102	5/17	2.00	2.00	0.00
103	5/16	2.00	2.00	0.00
104	5/15	2.00	2.00	0.00
105	111	5.00	5.33	0.33
106	112	6.00	6.67	0.67
107	155	37.00	36.44	–0.56
108	5/14	2.00	2.00	0.00
109	5/13	2.00	2.00	0.00
110	5/12	2.00	2.00	0.00
111	5/11	2.00	2.00	0.00
112	108	5.00	5.33	0.33
113	109	5.00	5.33	0.33
114	110	5.00	5.33	0.33
115	5/10	2.00	2.00	0.00
116	107	5.00	5.33	0.33
117	106	6.00	6.00	0.00
118	5/9	2.00	2.00	0.00
119	5/8	2.00	2.00	0.00
120	5/7	2.00	2.00	0.00
121	5/6	2.00	2.00	0.00
122	97	5.00	4.67	–0.33
123	98	4.00	5.10	1.10
124	99	5.00	5.33	0.33
125	100	5.00	6.22	1.22
126	101	5.00	5.33	0.33
127	102	5.00	5.33	0.33
128	103	5.00	5.33	0.33
129	105	5.00	5.33	0.33
130	96	4.00	4.00	0.00
131	104	5.00	5.33	0.33
132	5/5	2.00	2.00	0.00
133	84	10.00	10.00	0.00
134	67	10.00	8.67	–1.33
135	87	10.00	6.67	–3.33
136	24/3	5.00	5.33	0.33
137	24/4	5.00	5.33	0.33
138	24/5	5.00	5.33	0.33
139	23/2	2.00	1.90	–0.10

(Contd.)

Table 7.2 (*Contd.*)

S. No	*Khasra_No RoR*	*RoR_Area_M*	*Vec_Area_M*	*Area_Diff*
140	23/3	2.00	2.00	0.00
141	88	13.00	13.00	0.00
142	21	80.00	80.83	0.83
143	19/2	37.00	36.64	–0.36
144	93	5.00	4.46	–0.54
145	94	5.00	7.33	2.33
146	95	5.00	6.16	1.16
147	62	430.00	502.81	72.81
148	151	4.00	3.43	–0.57
149	61	984.00	765.45	–218.55
150	152	4.00	4.00	0.00
151	153	3.00	3.83	0.83
152	154	7.00	10.43	3.43
153	55	40.00	39.99	–0.01
154	150	3.00	2.67	–0.33
155	58	43.00	34.62	–8.38
156	149	34.00	34.33	0.33
157	148	8.00	8.22	0.22
158	147	89.00	89.00	0.00
159	10	150.00	160.00	10.00
	Total	8898.00	8838.81	–59.20

Table 7.3: Roads

S. No	*Khasra_ROR*	*Poly_Type*	*diff. (Marla)*
1	165	Road	140
2	168	Road	1
3	57	Road	–4
4	59	Road	–22
5	160	Cart_Track	–1
6	158	Road	21
7	161	Cart_Track	–2
8	156	Cart_Track	–20
9	56	Road	–108
.10	60	Road	46
11	55/1	Road	–6
12	82	Road	1
13	155	Road	–1
14	67	Road	–1
15	87	Road	–3
16	62	Road	72
17	151	Cart_Track	–1
18	61	Road	–215
19	153	Cart_Track	1
20	154	Road	3
21	58	Road	–8

Table 7.4: Difference in Area (>1 Marla) between Musavi and RoR and Their Analysis.

S. No	*Murraba No*	*Computerrized*			*RoR*		*difff >1 marla*	*Remarks*
		Kiila No	*K*	*M*	*K*	*M*	*M*	
1	*2*	*3*	*4*	*5*	*6*	*7*	*8*	*9*
1	0	120	0	5	0	3	2	This Khasra is adjacent to Laldora area. RoR shows 5 Marlas. Field Book shows 4 Marla area while the computed area of Musavi was 3 Marlas. Jamabandi after Misal Hakiyat shows 5 Marlas. Apparently, 1 Marla got added on after Misal haqiyat which needs to be looked into by DLR
2	0	121	1	3	0	19	4	This Khasra is adjacent to Laldora area. There is discrepancy between Map area and other records. According to our revenue expert there is calculation error in the fieldbook
3	0	123	1	0	1	2	−2	This Khasra is adjacent to Laldora area. RoR shows 5 Marlas. Field Book shows 4 Marla area while the computed area of Musavi was 3 Marlas. Jamabandi after Misal Hakiyat shows 5 Marlas. Apparently, 1 Marla got added on after Misal haqiyat which needs to be looked into by DLR
4	0	125	1	0	0	18	2	This Khasra is adjacent to Laldora area. There is discrepancy between Map area and other records. The deviation could be because of ignoring the decimal value.
5	0	126	0	19	1	2	−3	This Khasra is adjacent to Laldora area. There is discrepancy between Map area and other records. The deviation could be because of ignoring the decimal value.
6	0	133	0	3	0	5	−2	This Khasra is adjacent to Laldora area. There is discrepancy between Map area and other records. The deviation could be because of ignoring the decimal value.
7	0	154	0	7	0	11	−4	Adjacent Khasras also have a problem. There is discrepancy amongst different records, Musavi not legible.
8	0	156	1	15	0	15	20	Road. Fieldbook and Misal Hakiyat area matches with the Musavi area. Jamabandi area changed after Misal Hakiyat.

(Contd.)

Table 7.4 (*Contd.*)

1	2	3	4	5	6	7	8	9
9	0	158	3	17	4	18	–21	Road. Fieldbook, Misal Hakiyat, Jamabandi matches. Musavi area does not match.
10	0	161	0	9	0	7	2	Road. Near Laldora.
11	0	165	0	15	7	15	–140	Road. Fieldbook and Musavi area have a difference of 20 Marlas. However, Jamabandi and RoR area have over 120 Kanals difference. Looks like a data entry error in Jamabandi.
12	0	50	115	15	13	2	53	Laldora area.
13	0	55/1	6	14	6	8	6	Road. Difference between Musavi and other records.
14	0	56	22	13	17	5	108	Road. Difference between Musavi and other records. Probably, area of Khasra 23_24 belonging to Panchayat was demarcated for road.
15	0	57	15	19	15	15	4	Road. Difference between Musavi and other records.
16	0	58	2	3	1	15	8	Road. Difference between Musavi and other records.
17	0	59	7	10	6	9	21	Road. Difference between Musavi and other records. Check Khasras 25_22/1 and 26_17.
18	0	60	15	18	18	4	–46	Road. Difference between Musavi and other records. Check Khasra 14_24 belonging to school.
19	0	61	49	4	38	2	222	Road. Difference between Musavi and other records. Check Khasras 8_24 and 7_19.
20	0	62	21	10	25	2	–72	Road. Differe nce between Musavi and other records. Check the strip of Khasras along Abadi on the South and South East. The area should decrease here and the Abadi area should increase.
21	0	69	1	0	0	15	5	Khasra adjacent to road. Check for possible mistake on the road dimension as the road has excess area
22	0	72	1	0	1	4	–4	Adjacent to Laldora. Musavi not matching with other records. Excess area for owner.
23	0	73	1	0	1	2	–2	Near Laldora. Musavi area matching with Fieldbook. Varying with Jamabandi.
24	0	76	1	0	1	2	–2	Near Laldora. Musavi area not matching with other records.
25	0	85	1	0	0	18	2	Near Laldora. Musavi area not matching with other records.
26	0	87	0	10	0	7	3	Near Laldora. Musavi area matching with Fieldbook. Varying with Jamabandi.

(*Contd.*)

Table 7.4 (*Contd.*)

1	2	3	4	5	6	7	8	8
27	0	94	0	5	0	7	–2	Near Laldora. Musavi area not matching with other records
28	14	12	7	5	6	19	6	Musavi area not matching with other records
29	14	18	9	16	10	1	–5	Near Laldora. Musavi area not matching with other records
30	14	24	11	19	11	10	9	Area varying in different records. Near Laldora
31	15	18	3	18	4	7	–9	Near Laldora. Musavi area not matching with other records. Roads on 2 sides. North and North-east. Check on them.
32	15	21/2	3	0	2	17	3	Near Laldora. Musavi area not matching with other records. Roads on 2 sides. North and North West. Check on them.
33	15	22	7	14	7	11	3	Near Laldora. Musavi area not matching with other records. Roads on 2 sides. North and North West. Check on them.
34	16	13/1	4	0	0	18	62	Problem with Sub-Division
35	16	13/2	4	0	7	2	–62	Problem with Sub-Division
36	16	19	7	8	3	0	88	Musavi area matching with Fieldbook. Differs with other records. Possible encroachment and manipulation of record as the adjacent land belongs to panchayat
37	16	3/1/1	1	14	2	11	–17	Although on the boundary of village the problem could be with sub-division as the adjacent Khasras have a matching variation
38	16	3/1/2	1	13	1	18	–5	Although on the boundary of village the problem could be with sub-division as the adjacent Khasras have a matching variation
39	16	3/2	3	15	2	13	22	Although on the boundaryof village the problem could be with sub-division as the adjacent Khasras have a matching variation
40	16	6	8	3	7	20	3	On the Village boundary. Musavi area not matching with other records. Possible problem with rounding off of area computations or length as the polygon is zag zag.

(*Contd.*)

Table 7.4 (*Contd.*)

1	2	3	4	5	6	7	8	9
41	17	3	5	9	5	5	4	On the Village boundary. Musavi area not matching with other records. Possible problem with rounding off of area computations or length as the polygon.
42	17	8	4	1	4	8	–7	On the Village boundary. Musavi area not matching with other records. Possible problem with rounding off of area computations or length as the polygon is zag zag.
43	2	8	2	13	3	13	–20	Musavi area matching with Fieldbook. Differs with Jamabandi. On the boundary.
44	20	14	4	4	3	11	13	All records are varying. Possible error in field book. On the village boundary.
45	20	18	6	18	6	20	–2	Musavi area differs with other records.
46	21	2	5	9	5	7	2	Musavi and RoR are closely matching but varying with Fieldbook. Possible error in Fieldbook.
47	21	3	8	16	8	14	2	Musavi area differs with other records.
48	21	7	4	17	4	20	–3	Musavi area differs with other records.
49	21	8	8	4	8	2	2	Musavi area differs with other records.
50	23	24	8	0	9	16	–36	Musavi area matching with Fieldbook and Misal Haqiyat. Differs with other records.
51	23	25	4	4	4	2	2	Musavi area differs with other records.
52	25	14	7	18	8	0	–2	Musavi area matches with Fieldbook but differs with Jamabandi. Possible road demarcation not reflected in Musavi.
53	26	17	3	18	4	12	–14	Musavi area differs with other records. On the village boundary.
54	27	13	8	0	8	8	–8	Musavi area differs with other records. Could be a Fieldbook error.
55	27	16	6	12	8	0	–28	Musavi area matches with Fieldbook but differs with Jamabandi. Possible road demaracation not reflected in Musavi.
56	29	11	7	16	7	19	–3	Mismatch between Musavi, RoR and Jamabandi.
57	29	3	8	5	11	0	–55	Musavi and Fieldbook, First Jamabandi area matches closely. Lot of variance with RoR.
58	29	5	10	7	8	0	47	Musavi matches with Jamabandi but large variation with RoR.

(*Contd.*)

Table 7.4 (*Contd.*)

1	2	3	4	5	6	7	8	9
59	29	8/2	1	7	1	16	–9	Neighbouring S. Nos. 29_9/1. 29_3 also belong to the same Owner and the Musavi area exceeds the RoR area. Fieldbook and Misal Hakiyat sheets are missing. Need to look into it in totality.
60	29	9/1	5	11	7	4	–33	Neighbouring S. Nos. 29_8/2, 29_3 also belong to the same Owner and the Musavi area exceeds the RoR area. Fieldbook and Misal Hakiyat sheets are missing. Need to look into it in totality.
61	29	9/2	3	10	3	12	–2	Musavi area differs with RoR. Fieldbook data is missing. The difference could be because of ignoring the decimal.
62	3	25	2	13	3	6	–13	Musavi area differs with other records. On the village boundary.
63	30	5/1	5	16	5	12	4	Musavi area differs with other records. Road on one side. However, it could be wrong depiction of dimensions.
64	31	1	7	4	5	15	29	Musavi area differs with other records. Roads on two sides. However, the area adjacent to this Khasra also belongs to the same owner. It is possible that it could be a updating error.
65	31	6	8	0	9	13	–33	Musavi area matches with other records except RoR. Some changes could have occurred which is not reflected in Musavi.
66	32	16	7	10	7	13	–3	Musavi area differs with other records. On the village boundary.
67	32	19/1	1	15	2	14	–19	Different records depict different areas. Could be a problem with sub-division updating.
68	32	19/2	3	9	1	19	30	Different records depict different areas. Could be a problem with sub-division updating.
69	34	4	1	18	2	3	–5	Musavi area differs with other records. On the village boundary and road on one side. May be a problem with road dimensions.
70	35	24	7	4	7	14	–10	Musavi area matches with other records except RoR. Some changes could have occurred which is not reflected in Musavi.
71	37	22/1/2	1	4	1	14	–10	Musavi and Fieldbook, First Jamabandi area matches closely. Variance with Jamabandi and RoR areas.
72	38	9/1	9	0	9	3	–3	Musavi area differs with other records. The adjacent Khasra has an excess area of 3 Marlas. May be, sub-division updating error

(*Contd.*)

Table 7.4 (*Contd.*)

1	2	3	4	5	6	7	8	9
73	38	9/2	2	0	1	17	3	Musavi area differs with other records. The adjacent Khasra has less area of 3 Marlas. May be, sub-division updating error.
74	39	2	6	11	7	9	–18	Musavi area differs with other records. On the village boundary. Could be an area calculation error.
75	39	20	8	7	8	2	5	Musavi area differs with other records. On the village boundary. Could be an area calculation error.
76	39	23	4	0	3	18	2	Musavi area differs with other records.
77	4	21	3	5	3	13	–8	Musavi area differs with other records. On the village boundary. Could be an area calculation error.
78	47	11	8	0	7	11	9	Musavi and Fieldbook area matches. Variance with Jamabandi and RoR areas. Adjacent Khasra belongs to the same owner. If clubbed the variance is only 2 Marlas.
79	47	12	8	0	8	7	–7	Musavi and Fieldbook area matches. Variance with Jamabandi and RoR areas. Adjacent Khasra belongs to the same owner. If clubbed the variance is only 2 Marlas.
80	49	4/1	3	5	2	18	7	Musavi area differs with other records. On the village boundary. Possible error in depiction of dimension on Musavi.
81	6	15	2	3	2	6	–3	Musavi area differs with other records. On the village boundary. Possible error in depiction of dimension on Musavi.
82	6	16	2	12	2	15	–3	Musavi area differs with other records. On the village boundary. Possible error in depiction of dimension on Musavi.
83	6	17/1	3	13	3	16	–3	Musavi area differs with other records. Possible error in depiction of dimension on Musavi.
84	6	19/1	3	12	3	9	3	Musavi area differs with other records. Adjacent to road.
85	6	9/2	1	2	0	18	4	Musavi area differs with other records. The adjacent Khasra has an excess area of 4 Marlas. May be, sub-division updating error
86	6	9/3	0	11	0	15	–4	Musavi area differs with other records. The adjacent Khasra has an excess area of 4 Marlas. May be, sub-division updating error

(*Contd.*)

Table 7.4 (*Contd.*)

1	2	3	4	5	6	7	8	9
87	7	11	7	6	7	2	4	Different records depict different areas. There is a general problem in Khasras around this area.
88	7	18	4	11	4	15	–4	Different records depict different areas. There is a general problem in Khasras around this area. Also, road on one side.
89	7	19	6	4	6	13	–9	Different records depict different areas. There is a general problem in Khasras around this area. Also, road on one side.
90	7	27/2	4	7	3	20	7	Different records depict different areas. There is a general problem in Khasras around this area.
91	7	27/3	4	10	4	8	2	Musavi area differs with other records. Error in Fieldbook.
92	8	15	10	16	10	18	–2	Musavi area differs with other records. However, closely matches with RoR. Fieldbook and Jamabandi areas are in great variance. There is a general problem in Khasras around this area.
93	8	21	3	16	3	4	12	Musavi area differs with other records. Road on one side. Could be a dimension depiction issue or area calculation issue
94	8	24	3	2	5	2	–40	Musavi area differs with other records. Road on one side. Could be a dimension depiction issue or area calculation issue
95	8	27	9	11	9	14	–3	Different records depict different areas. There is a general problem in Khasras around this area.
96	8	28	8	12	8	4	8	Different records depict different areas. There is a general problem in Khasras around this area.
97	9	24/2	2	2	2	5	–3	Musavi area differs with other records. Road on one side. Could be a dimension depiction issue or area calculation issue.
98	9	8	8	0	8	15	–15	Different records depict different areas. Khasra on the village boundary. Dimension or area calculation error.
99	48	15	—	—	7	3	–143	Data not available in RoR. Matches with Fieldbook with 3 Marla difference.
100	48	6/1	—	—	4	12	–92	Data not available in RoR. Matches with Fieldbook.

(*Contd.*)

Table 7.4 (*Contd.*)

1	2	3	4	5	6	7	8	9
101	36	15/12	—	—	0	9	–9	Data not available in RoR. Lot of variance with fieldbook. Although a cart track, FB shows 1 acre area.
102	38	15/2	—	—	3	2	–62	Data not available in RoR. Matches with Fieldbook.
103	38	15/1	—	—	4	18	–98	Data not available in RoR. Matches with Fieldbook.
104	32	12	—	—	10	4	–204	Data not available in RoR. Matches with Fieldbook.
105	33	2/1	—	—	3	13	–73	Data not available in RoR. Matches with Fieldbook with 1 Marla difference.
106	23	19	—	—	8	0	–160	Data not available in RoR. Matches with Fieldbook.
107	22	17/1/1	—	—	3	4	–64	Data not available in RoR. Lot of variance with fieldbook. Possible non-updating of record in Fieldbook as the composite area of the adjacent Khasras does not match with Kila area.
108	17	13/1	—	—	0	4	–4	Data not available in RoR. Lot of variance with fieldbook. Possible non-updating of record in Fieldbook.
109	17	13/2	—	—	6	13	–133	Data not available in RoR. Lot of variance with fieldbook. Possible non-updating of record in Fieldbook.
110	14	8/7	—	—	8	18	–178	Data not available in RoR. Matches with Fieldbook with 5 Marla difference.
111	15	5/2	—	—	0	2	–2	Data not available in RoR. Matches with Fieldbook.
112	6	24/2	—	—	0	5	–5	Data not available in RoR. Matches with Fieldbook.
113	6	24/6	—	—	0	5	–5	Data not available in RoR. Matches with Fieldbook.
114	6	23/6	—	—	0	2	–2	Data not available in RoR. Matches with Fieldbook.

Table 7.5: Burgi to Burgi Differences amongst Adjoining Villages of Pilot Villages

Village/HB No.:	*Berheri / 27*		*Tehsil: JAGADHARI*		*District: YAMUNANAGAR*		*Vector Length of Berheri Village (Burji to Burji)*
Sl. No.	*Burji Number*		*Length (in Karam)*	*Adjacent Village Name*	*Length (in Karam)*	*Difference In Length (If Any)*	
	From	*To*					
1	*2*	*3*	*4*	*5*	*6*	*7*	*8*
97	BB0007	CB0093	136	Palewala	136	0	137
98	CB0093	CB0094	29	Palewala	29	0	29
99	CB0094	CB0095	33	Palewala	33	0	33
100	CB0095	CB0096	37	Palewala	37	0	37
101	CB0096	BB001	85	Palewala	85	0	85
1	CB001	CB002	67	Palewala	67	0	67
2	CB002	CB003	22	Palewala	22	0	22
3	CB003	CB004	43	Palewala	43	0	43
4	CB004	CB005	40	Palewala	40	0	40
5	CB005	BB001	8	Palewala	8	0	8
Differences Between Berheri and Palewala Village			**500**		**500**	**0**	**502**
	BB001	CB006	30	Mohari	30	0	31
6	CB006	CB007	75	Mohari	75	0	76
7	CB007	CB008	4	Mohari	4	0	4
8	CB008	CB009	43	Mohari	43	0	43
9	CB009	CB0010	67	Mohari	67	0	67
10	CB0010	CB0011	43	Mohari	43	0	43
11	CB0011	CB0012	104	Mohari	104	0	103
12	CB0012	CB0013	34	Mohari	34	0	34
13	CB0013	CB0014	65	Mohari	65	0	65
14	CB0014	CB0015	24	Mohari	24	0	25

(Contd.)

Table 7.5: (*Contd.*)

1	*2*	*3*	*4*	*5*	*6*	*7*	*8*
15	CB0015	CB0016	41	Mohari	41	0	41
16	CB0016	CB0017	3	Mohari	3	0	4
17	CB0017	CB0018	37	Mohari	37	0	37
18	CB0018	CB0019	48	Mohari	48	0	49
19	CB0019	CB0020	26	Mohari	26	0	26
20	CB0020	CB0021	76	Mohari	76	0	76
21	CB0021	CB0022	34	Mohari	34	0	34
22	CB0022	CB0023	112	Mohari	112	0	111
23	CB0023	CB0024	15	Mohari	15	0	15
24	CB0024	CB0025	50	Mohari	50	0	50
25	CB0025	BB002	54	Mohari	54	0	54
Differences Between Berheri and Mohari Village			**985**		**985**	**0**	**989**
	BB002	CB0026	94	Jathlana	94	0	94
26	CB0026	CB0027	31	Jathlana	31	0	31
27	CB0027	CB0028	53	Jathlana	53	0	52
28	CB0028	CB0029	39	Jathlana	39	0	39
29	CB0029	CB0030	16	Jathlana	16	0	16
30	CB0030	CB0031	24	Jathlana	24	0	23
31	CB0031	BB003	72	Jathlana	72	0	72
Differences Between Berheri and Jathlana Village			**329**		**329**	**0**	**328**
	BB003	CB0032	116	Lalchppar	116	0	118
32	CB0032	CB0033	13	Lalchppar	13	0	13
33	CB0033	CB0034	17	Lalchppar	17	0	17
34	CB0034	CB0035	22	Lalchppar	22	0	22
35	CB0035	CB0036	12	Lalchppar	12	0	12
36	CB0036	CB0037	38	Lalchppar	38	0	38

(*Contd.*)

Table 7.5: *(Contd.)*

1	*2*	*3*	*4*	*5*	*6*	*7*	*8*
37	CB0037	CB0038	20	Lalchppar	20	0	20
38	CB0038	CB0039	46	Lalchppar	46	0	46
39	CB0039	CB0040	29	Lalchppar	29	0	29
40	CB0040	CB0041	9	Lalchppar	9	0	9
41	CB0041	CB0042	28	Lalchppar	28	0	30
42	CB0042	CB0043	69	Lalchppar	69	0	69
43	CB0043	CB0044	38	Lalchppar	38	0	41
44	CB0044	CB0045	56	Lalchppar	56	0	56
45	CB0045	CB0046	185	Lalchppar	185	0	185
46	CB0046	CB0047	144	Lalchppar	144	0	143
47	CB0047	CB0048	117	Lalchppar	117	0	117
48	CB0048	CB0049	11	Lalchppar	11	0	11
49	CB0049	CB0050	32	Lalchppar	32	0	32
50	CB0050	CB0051	11	Lalchppar	11	0	11
51	CB0051	BB004	81	Lalchppar	81	0	81
Differences Between Berheri and Lalchppar Village			**1094**		**1094**	**0**	**1100**
52	BB004	CB0052	39	Sandhli	39	0	39
53	CB0052	CB0053	8	Sandhli	8	0	8
54	CB0053	CB0054	84	Sandhli	84	0	83
55	CB0054	CB0055	133	Sandhli	133	0	134
56	CB0055	BB005	203	Sandhli	203	0	203
Differences Between Berheri and Sandhli Village			**467**		**467**	**0**	**467**
58	BB005	CB0056	35	Mandhar	35	0	35
59	CB0056	CB0057	26	Mandhar	26	0	26
60	CB0057	CB0058	29	Mandhar	29	0	29
61	CB0058	CB0059	14	Mandhar	14	0	14

(Contd.)

Table 7.5: (*Contd.*)

1	*2*	*3*	*4*	*5*	*6*	*7*	*8*
62	CB0059	CB0060	25	Mandhar	25	0	25
63	CB0060	CB0061	47	Mandhar	47	0	47
64	CB0061	CB0062	35	Mandhar	35	0	36
65	CB0062	CB0063	67	Mandhar	67	0	66
66	CB0063	CB0064	96	Mandhar	96	0	96
67	CB0064	CB0065	30	Mandhar	30	0	30
68	CB0065	CB0066	30	Mandhar	30	0	30
69	CB0066	CB0067	44	Mandhar	44	0	45
70	CB0067	CB0068	47	Mandhar	47	0	47
71	CB0068	CB0069	25	Mandhar	25	0	25
72	CB0069	CB0070	30	Mandhar	30	0	30
73	CB0070	CB0071	16	Mandhar	16	0	16
74	CB0071	BB006	4	Mandhar	4	0	4
Differences Between Berheri and Mandhar Village			**600**		**600**	**0**	**601**
75	BB006	CB0072	76	Madhubans	76	0	76
76	CB0072	CB0073	29	Madhubans	29	0	29
77	CB0073	CB0074	41	Madhubans	41	0	41
78	CB0074	CB0075	18	Madhubans	18	0	18
79	CB0075	CB0076	117	Madhubans	116	1	117
80	CB0076	CB0077	12	Madhubans	12	0	13
81	CB0077	CB0078	23	Madhubans	23	0	22
82	CB0078	CB0079	35	Madhubans	35	0	35
83	CB0079	CB0080	22	Madhubans	22	0	22
84	CB0080	CB0081	47	Madhubans	47	0	47
85	CB0081	CB0082	34	Madhubans	34	0	34
86	CB0082	CB0083	13	Madhubans	13	0	13

(*Contd.*)

Table 7.5: *(Contd.)*

1	*2*	*3*	*4*	*5*	*6*	*7*	*8*
87	CB0083	CB0084	63	Madhubans	63	0	63
88	CB0084	CB0085	20	Madhubans	21	–1	20
89	CB0085	CB0086	31	Madhubans	31	0	31
90	CB0086	CB0087	66	Madhubans	66	0	65
91	CB0087	CB0088	76	Madhubans	76	0	76
92	CB0088	CB0089	30	Madhubans	30	0	30
93	CB0089	CB0090	21	Madhubans	21	0	21
94	CB0090	CB0091	23	Madhubans	24	–1	22
95	CB0091	CB0092	14	Madhubans	14	0	14
96	CB0092	BB0007	22	Madhubans	22	0	22
Differences Between Berheri and Madhubans Village			**833**		**834**	**–1**	**833**
Total Differences Between Berheri and Adjoining Villages Burjis					**(–1) Karam**		**13**
Total Area in Karam			**4808**		**4809**		**4821**
Area Difference between the Berheri recorded boundary and the Vector Boundary in Feet							**71 F**
Area Difference between the Berheri recorded boundary and the Vector Boundary in Metres							**22 M**

Table 7.6: Data not Available in RoR for the following Khasras in Musavi

S. No	*Murabba_ No.*	*Khasra_ No.*	*Vector_ Kanal*	*Vector_ Marla*	*Vector_ KNL&MRL*
1	48	15	7	3	143
2	48	6/1	4	12	92
3	36	15/12	0	9	9
4	38	15/2	3	2	62
5	38	15/1	4	18	98
6	32	12	10	4	204
7	33	2/1	3	12	72
8	23	19	8	—	160
9	22	17/1/1	3	4	64
10	17	13/1	0	4	4
11	17	13/2	6	13	133
12	14	8/7	8	18	178
13	15	5/2	0	2	2
14	6	24/2	0	5	5
15	6	24/6	0	5	5
16	6	23/6	0	2	2
17	8	5555	1	1	21

Chapter – 8

Bhoomi: An Online Land Records Management System

Rajeev Chawla, A. Venkatesan, and *N.R. Samartharam*

Note to Practitioners

1. Change management strategies are very critical for implementation of any e-Governance project. In BHOOMI also change management played vital role. Aspects like clarity in vision and continuous and gradual improvement discussed in the latter part of this case study played important role in making BHOOMI sustainable.
2. Ownership from the top government functionaries in the form of Chief Minister, Revenue Minister, and Principal Secretary to Revenue Department really comes handy while implementing such mammoth project.
3. Concentrated and continuous efforts in training and capacity building are significant factors in making project sustainable.
4. Government Process Re-engineering in terms of timely amendments to relevant acts and change in practices and procedure in computerized system help in realizing desired objectives.
5. Continuation of project champion for complete period of rollout and single command for all land records management activity in terms of integration and reforms go long way in achieving Total Land Records Management.

Project Context—Background

Land has been very important aspect of the mankind from the time immemorial. Earliest evidences of land records system dates back to erstwhile kingdoms ruled ancient India. Records pertaining to land with measurements were critical for collection of revenue. Land revenue was the major source of income to the state. One can notice from the records that there existed a quite refined land records system during the period of Sher Shah Suri (1534-1545). Todarmal one of the ministers of the King Akbar (1556-1605) is known for his skills in land administration. Land records administration has undergone lot of evolution process since Mughal period. Whatever system which is being used today in India has strong influence of British rule on Moghal land administration system. Land revenue system introduced by British in India had three broad types, namely "*Permanent Settlement*", "*Mahalwari Settlement*" and "*Ryotwari System*" with each system incorporating the elements of the preceding agrarian structure. The "*Ryotwari System*" prevailed in southern part of the country including Karnataka, wherein individual cultivators were recognized as owners of the land. During British period and post-independent period till 1964, there were different land revenue code and rules used across Karnataka. Land records administration was based on "*Mysore Land Revenue Code and Rules 1888*" in old Mysore area which was ruled by erstwhile Mysore kings. In Bombay Karnataka region it was based on "*Manual of Revenue Accounts of the Villages, Taluks and Districts of Bombay States, by F.G.H. Anderson I.C.S.*". The areas which came from Madras Presidency and Hyderabad-Karnataka region largely followed similar system based on "*Madras Board of Revenue Act 1883*". The district of Coorg was following *"Coorg Land Revenue Regulation 1899"* though it was under Madras presidency. *"Karnataka Land Revenue Act 1964"* repealed all the above acts to give consolidated common land revenue act to State of Karnataka.

A good land records system irrespective of whether it is electronic or manual is a necessity for any harmonious and progressive society. Over the years, significance of error-free, tamper proof and effortlessly accessible land records has been the focus in India. It is estimated that about 30 million cases are pending across various courts in the country and astoundingly eighty per cent of them are civil cases and greater part of these civil cases are in some way or the other, land related disputes.

Project Context—Pre-implementation Scenario

In the manual system RoRs (called as RTC—Record of Rights, Tenancy and Cultivation/crops in Karnataka) were maintained by 9,000 Village Accountants (a field level functionary at the bottom of the revenue hierarchy) across the State. Jurisdiction of Village Accountant is called as Village Accountant Circle and on an average it spans over 3-4 villages depending on area, number of RTCs, area under irrigation, etc., Village Accountant's duty is executive in nature, he/she is supposed to move around villages to provide service to citizen at their doorstep. There is no fixed office where he/she needs to sit and perform his/her duties, making life of citizen difficult to locate him/her in urgency. Village Accountants are supervised by official called Revenue Inspector at sub-taluk level (these are called as *hoblis* in Karnataka), a typical *hobli* would consist of 8-10 village accountant circles. Any revenue official equal to the rank of Revenue Inspector or above is authorized to accept the changes to RTCs in the form of mutations. This essentially means that Revenue Inspector is lowest cadre official who can approve mutation and he is the person who is authorized to take decision on major chunk of mutations. Inheritance mutation cases were handled by RRT—*sheristedar* an official who is one rank above Revenue Inspector and operates from Taluk office. Like Village Accountant, Revenue Inspector also works as executive and he does not have any fixed office. Land Records administration activities of taluka are supervised by Tahsildar, who has powers to hear objections rose against mutations and accordingly pass orders. Assistant Commissioner at sub-division level and Deputy Commissioner at district level also monitor the land administration activity with powers to hear appeal against decisions of their subordinate officers.

All the RTCs pertaining to a revenue village used to be maintained in the form of hardbound book in the ascending order of survey number. These books were rewritten once in five years as space available for writing crop information used to suffice only for that period. Any changes happening due to various types of mutations

(The word *Mutation* is used to represent the act or process of changing or altering contents of RTC other than crop information) due to sale, inheritance, government orders, court order, pledge, release, *phodi*, etc., were recorded on the existing RTC by rounding off the old contents and writing new details on the same sheet, only exception being *phodi* transaction, where existing RTC used to be cancelled and new RTCs were written in the end using blank formats bound for the purpose. These bound RTC books were maintained in duplicate, one copy was kept in taluk office which was called original and duplicate copy for use to be with Village Accountant. Village Accountant was expected to carry out changes resulting out of mutations and crop updation both in original RTC kept in taluk office and duplicate which was with him.

Over the years due non-updation of taluk office copy of RTC which was supposed to be original became stale and duplicate which Village Accountant use to keep with him became up-to-date and thereby resulting into *defacto* original. This resulted in a scenario with no alternative for citizens except to approach Village Accountant for copy of their RTC. The distribution mechanism by some of the Village Accountants was based on the background of the person asking for the document. Rich and politically powerful people were served in the place convenient to them with or without fee. Poor and downtrodden has to make multiple visits and spend relatively more money to get the copy of their record of rights. Copies were issued to the convenience of the citizen depending on the need, of course for a huge consideration. Changing crop grown from food crop to commercial crop to enable citizen to avail farm credit was a common practice in some areas. Some unscrupulous elements in revenue department use to change land type from wet/garden to dry while issuing certified copies to enable citizen to enjoy cut in stamp duty during registration. Many a time courts were misguided by issuing wrong certified copies for wrong motives.

When it comes to procedure followed in carrying out mutations, backdated mutations were very common. Mutations carried out without serving notices resulted in spurt in number of cases in revenue courts. Preferential treatment for mutation applications and J-slips (Intimation slip informing Tahsildar about the registration event) from sub-registrars was well known thing. Mutation transactions on J-slips, supplementary *akarbhand*, government orders such as alienation and acquisition were never attempted voluntarily. Discretion of the officials on what to attempt and what not to attempt was at its peak. There were instances where mutation for J-slip of 1930 was initiated after BHOOMI. Above all non-incorporation of mutation entries to RTC was greatest problem. In many cases mutations were approved by Revenue Inspectors and entries which were supposed to have been done by Village Accountant were not carried out due to extraneous reasons.

All the factors and the problems explained above aggravated over the years and resulted into a land records management system which are:

- Opaque;
- Prone to manipulations;
- Harassment and Extortion;
- Delay in delivery of land records;
- Cumbersome mutation process;
- Lack of timely data for planning purpose;
- Cumbersome crop loan mechanism; and
- Delay in disposal of civil litigations.

Keeping inherent problems explained in above paragraphs, right from the First Five Year Plan, planners have been advocating proper maintenance of land records as the basis for good administration, aimed at social justice through better implementation of rural development programmes. This was reiterated in the Second and Third Plans. The Sixth Plan had envisaged the completion and updation of land records from 1980 to 1985. Centrally-sponsored scheme on computerization of land records was started in 1988-89 with 100 per cent financial assistance. A good land records management system will have the following benefits:

- Increase in land value;
- Increase in demand and supply of credits;
- Interest rates would be cheaper;
- Incidence of land transactions increase with good land records;

- Increase in investment on farm inputs such as seeds and chemical fertilizers;
- Yield per unit area increase;
- Increase in cultivated area;
- Farmers switching cropping systems to horticulture crops;
- Reduction on poverty in rural areas; and
- Reduction in land disputes strengthened communities.

Project Description

BHOOMI is the project conceived and implemented by Government of Karnataka with National Informatics Centre as Technical partner based on the funding received from Government of India under CLR (Computerization of Land Records), now renamed as NLRMP (National Land Records Modernization Programme) to have good land records management system which in itself, a good example of a paradigm shift from department centric approach to citizen centric approach and process oriented to service oriented. Under BHOOMI project about 20 million RTCs of 6.7 million farmers of Karnataka State have been digitized. All these records have been verified and certified by the revenue authorities before making digital database as the only source of land records. Manual records have been invalidated through amendment to Karnataka Land Revenue Rules 1964. Any changes to record of rights will happen through online mutation application (BHOOMI) only. Changes to RTCs may be due to sale, partition, inheritance, pledge, release, government order, court order, *phodi*, etc. Online mutation application handles all type of mutation so that no manual intervention in updating land records database. Workflow based user-friendly software has been designed and developed by National Informatics Centre, Bangalore has bio-metric authentication to take care of non-repudiation and scanning interface to scan important documents as evidence for electronic transaction. PKI (Public Key Infrastructure) enabled BHOOMI database and application with integration of digital signatures is in place to adhere to IT Act 2000. On an average BHOOMI software handles more than 0.1 million mutation a month. About 9,000 Village Accountants, 800 Revenue Inspectors, 203 Sheristedars, 1000 Operators-cum-Village Accountants, 203 Tahsildars and 52 Assistant Commissioners work on the system based on clear-cut roles and privileges given to them.

Initially BHOOMI was implemented in 203 locations in phased manner from 1999 to 2002 as stand-alone system with local server and clients. In 2005 a V-SAT based network was establish to connect all the taluka servers to State Data Centre (SDC) and replication of BHOOMI database between taluk and SDC was put in place. The replicated data in SDC helps in disaster recovery activities in case data is lost due to any reason in the taluk server. Data in SDC is also used in publishing record of rights on web and developing interfaces with different stakeholders. After commissioning of KSWAN (Karnataka State Wide Area Network), V-SAT based network has been discarded and KSWAN is being used for communication between taluk and SDC.

Under BHOOMI project, maintenance of land records electronically was given first priority keeping all other interfaces as manual only. Initially all the registration events of registration department, farm credit declarations from Banks and co-operatives societies, notifications from land acquisition authorities, government and court orders, etc., were intimated to BHOOMI centre in paper form. BHOOMI project had employed "*Bottom Up*" approach in implementing e-Governance which is evolutionary/synthesizing model wherein start with simple activities and scale up fast to achieve larger objectives within the department individually. Once the individual systems are stable and sustainable then start integrating them to achieve ultimate objectives. Started as simple vanilla application to maintain land records, today BHOOMI has been enhanced in such way that it has electronic interfaces with other departments/wings. SDC database is being used to publish web services to departments/wings of government machinery to use the land records data and submit the transactions electronically so that mutations can be started automatically without human intervention. Details of these interfaces are discussed in length under the heading stakeholders.

Objectives of Project

The major objectives of the BHOOMI project were:

- Facilitating easy maintenance and prompt updating of land records.

- Making land records tamper-proof.
- Allowing farmers easy access to their records.
- Collating the information to construct a database regarding land revenue, cropping pattern, land-use, etc.
- Utilising the data for planning and for formulating development programmes.
- Integrating all the land related activities electronically with BHOOMI so as to achieve single source of RoR data and its updation with minimum or no human intervention.
- Enabling usage of this BHOOMI database by courts, banks, private organizations and companies, ISPs

BHOOMI has achieved most of these objectives listed above, only left out area is integrating with revenue and civil courts for which discussion are in progress so that court and BHOOMI can exchange data electronically and implement the court orders without delay.

Stakeholders in the Project

In this section, benefits to each of the stakeholders are discussed along with integration achieved to reduce the human discretion and to automate the process as much as possible.

Citizen: Citizen is the main stakeholder in BHOOMI project; BHOOMI is successful because its main beneficiary is citizen. Since citizen is involved directly, it becomes the responsibility of the government to provide continuous support and run the service unaffected once started in such G2C projects. This also matter a lot for the long time sustainability of the project, which is otherwise missing in G2G projects. Citizen has easy access to his/her record of rights without harassment and extortion; it used to take lot of time to get copy of the RoR before BHOOMI which is now available in 2 minutes. Alternatives enabled for accessing RoRs, such as rural tele centres and web publishing has reduced the dependency on government run taluk BHOOMI kiosks. Facility for citizen to file the mutation request has brought accountability in the system; acknowledgement with reference ID has simplified the process of tracking the request. A transparent online mutation system which has been put in place helps in avoiding pre-dated notices and overnight mutations which is a great sigh of relief to citizen. Availability of mutation status across counter and on web has provided much required confidence on BHOOMI project. Other benefits arising out of BHOOMI as by-products from other departments is described in the subsequent paragraphs of this section.

Registration Department: Registration department is one of the main consumers of the RoRs data as all the transactions to transfer the rights except inheritance transaction happens in Sub-Registrar Office (SRO), which is field level office under registration department. In the pre-BHOOMI scenario, registration department was only dependent on the documents submitted by the parties at the time of registration and there was no verification mechanism in place. There were instances where same piece of land had been sold more than once to different people. The vacuum of non-availability of data at the time of registration with respect to agricultural land has been filled by integrating BHOOMI and KAVERI (Registration software). Since 2006, BHOOMI is consuming the index data of transactions from KAVERI, now full and comprehensive integration has been achieved in Karnataka. The KAVERI system of the Stamps and Registration department now electronically talks to the BHOOMI system and ensures that the seller is indeed owner of the property as per the BHOOMI database. Soon after the transaction is over the extent in BHOOMI database of the seller gets decreased ensuring that he cannot sell more than what he owns or that he cannot sell the same piece of land to others.

Immediately after the transaction is over in Sub Registrar office a mutation transaction is started in BHOOMI system automatically as all the required data for starting of such transaction is available due to electronic linkages between the two systems. Citizens are also sent the SMS automatically as their transactions pass various phases/stages in the mutation process in BHOOMI system. The citizens, therefore, are more aware of what is happening on their lands and can also find out if any fraudulent transaction has taken place on their lands as they would certainly get SMS (just like credit card system).

Integration of BHOOMI and KAVERI has not only helped prospective buyers but has also gone a long way in enhancing the credibility of the registration department. Integration has also helped in stopping transactions on government lands or lands on which government restrictions exists. Further it blocks the transactions on the lands on which court has given order/stay. The integration also throws light on whether transaction needs pre-mutation sketch or not which has contributed a lot for reducing the disputes. Details about pre-mutation sketch and its importance in land administration are discussed under the heading survey department in this section.

Banks and Co-operative Institutions: If we analyze the different types of transactions happening in BHOOMI, more than fifty per cent of transactions pertain to pledge and release type. These are transactions wherein citizen has pledged his land to banks or co-operative institutions for getting farm credit or released the property after paying back the loan. On analyzing data for few taluks, it was clear that average time taken for data entry and initiating the transaction is maximum compared to any other stage. This is because, the volume of applications that are being received from these banks and co-operative institution for creating liabilities on the land. Keeping the volume of transactions and time taken for data entry of the same, Revenue Department has decided to integrate BHOOMI and Bank activities electronically. Facility has been enabled for banks and co-operative institutions to raise request for pledge or release over internet from the bank itself. The request will contain all the parameters to initiate the transaction in BHOOMI automatically. This has helped BHOOMI to reduce the pendency at data entry level and Banks are benefited because they can access the land records database for confirming ownership, extents owned by the farmer, other liabilities that farmer has and also monitor the status of requests raised by them. Banks and co-operatives institution will also become interested parties for all those land parcels for which they have disbursed farm credit.

Land Acquiring Bodies: Land acquisition is the process wherein agricultural and/or non-agricultural lands are acquired by government for various activities such as village/town extension, roads, culverts, reservoirs, canals, military camps, railways, industries, etc., Regular Assistant Commissioners in the revenue sub-divisions also act as Land Acquisition Officers along with special Land Acquisition Officers appointed by government, urban development authorities, Industrial development authorities, National Highways, etc., Keeping the volatile nature of land records, notifications are bound to go wrong keeping time required to collect the manual data and preparation of notification manually. Certain transactions which do not change the ownership such as *podi* keep happening even after notification resulting in change in key parameters itself. The paper based intimations from the land acquiring bodies were not getting implemented in land records system even though land records management was electronic in nature. It was up to the discretion of the officials to decide whether to initiate the transaction or not. In essence due to absence of direct electronic interface between the land acquisition officer and the administrator of record of right management, the notification of intended acquisition of land was not indicated on the record of right and most of the time there was a scope for changing the ownership, creation of liability on the land as well as there was an enormous scope for acquiring the same piece of land resulting in a huge loss to the State Exchequer in terms of paying compensation multiple times.

Electronic integration of BHOOMI and BHOOSWADEENA (An automated Land Acquisition System) has brought synchronization between two systems. Land acquiring bodies consume the land records data electronically to create notifications in a workflow based Land Acquisition System. After every stage of notification, BHOOMI is informed about the event by sending XML (eXtensible Markup Language) file, which in turn is used in BHOOMI to create a transaction automatically either to create liability entry about published notification or to replace the owner with the name of the acquiring body depending on the stage of activity. Timely updation of land records is helping the prospective buyers to take decisions as ground reality is reflected on records of rights with respect to acquisition proceedings that are in progress on a particular land parcel. Registration Department is also alerted before executing transaction as land acquisition notifications are updated on records of rights. On the other hand deserving landlords will be paid compensation without litigations. Computerization of Land Acquisition Offices has resulted in better management of land acquisition activity as a whole.

Survey Department: While BHOOMI was in place by the year 2000 there were fundamental problems with the spatial data management and synchronization of it with textual records. The process of *phodi* (division of bigger property into a smaller *hissas* on the ground) used to take lot of time after sale/inheritance transaction.

Many a time it used to take even 15-20 years. This resulted in a situation wherein 2-3 people were shown as owners in a single property record but their rights on the ground were not delineated clearly.

Karnataka Land Revenue Act was amended and it was mandated that a pre-mutation sketch is prerequisite for any sale transaction to happen. The pre-mutation sketch is created in the field using the services of licensed private surveyors on a basis of user charges and shows the area proposed for sale. This pre-mutation sketch is mandatory and no transaction in sub-registrar offices can happen without the sketch. Whole of this activity of issuing pre-mutation sketch has been simplified using software called 'MOJANI'. This mechanism has gone a long way to ensure that there are no disputes as the boundaries is known to the purchaser even before he purchases the land. The concept of private licensed surveyors has ensured that the public sector was not required to augment its own strength and there was no additional financial load on the system. Needless to say that MOJANI uses the BHOOMI database at the backend and is fully integrated with it for creation of any sketch. This scheme not only helps the Government and citizens but also help in employment of rural youths numbering about 3500.

Pre-mutation sketch which is mandatory for transfer of properties would be used during mutation process; new concept called Integrated Mutation Podi has been introduced to achieve this objective. Along with transfer of ownership, new RTC is also created with new land parcel details bringing synchronization of spatial and textual data of land records.

Other Departments: Different departments such as Department of Economics and Statistics, Agriculture, Horticulture, Forest, Agricultural related boards and corporations use BHOOMI data extensively. Crop details of RoRs are used by department of Economics and Statistics, Agriculture and Horticulture. Forest department uses RoRs data for protecting forest land.

Revenue Hierarchy and Administrators: MIS and Pendency reports have helped administrator and revenue hierarchy to monitor the land records maintenance activity effectively. Analysis provided through various reports is helping administrators to think new ways and means to improve the system.

Government: Prime objective of the BHOOMI project is being SMART (Simple, Moral, Accountable, Reliable and Transparent) land records management system and easy access to citizen, it has greater impact on the goodwill of the government. Successive governments in Karnataka have projected the project in their achievement list. Budget declarations with reference to the project indicate the popularity it brought to men in the power centre and resulting itself into political will for reforms and improvements

Services offered in BHOOMI

Through Taluk Kiosk

- RoRs (Current and old) — Both Current and old RoRs are available at the cost of Rs. Ten each. In case of PKI (Public Key Infrastructure) enabled taluks, digitally signed and bar coded RoRs without ink signature are being issued.
- Mutation extract — This is the document that gives the details about acquisition of right, certified copy of this document is available at the cost of Rs. Fifteen.
- RR5 and RR6 — Records of rights Form–5 and Form–6 which is commonly called as RR5 and RR6. These documents give history of transactions chronologically for a particular village and on a particular survey number for a given period respectively.
- Khatha Extract — All the land parcels owned by a person in a village will be listed in the Khatha extract.
- Mutation Status — A document which talks about stage at which the requested transaction is currently held up and from how many days.
- Scanned copies of served notice, mutation extract, Tippan– Scanned documents such as served notice having signatures of the interested parties, mutation order with signatures of Village Accountant and Revenue Inspector and scanned image of tippan (sketch of land parcel) are available for dissemination from taluk kiosk.

Through Rural Tele Centre

- RoRs (Current and old).
- Mutation extract.
- Mutation status.

Through Web (for View only)

- RoRs (Current and old).
- Mutation extract.
- RR5 and RR6.
- Mutation status.
- Scanned copies of served notice, mutation extract, Tippan.

Outcomes and Benefits realized by Implementing the Project

BHOOMI has brought in whole lot of changes in the way by which land records were administered. Manual records have been transformed into computerized ones. Transparency, easy accessibility, tamper proof, harassment free and citizen friendly kiosk for dissemination of RoRs are basic objectives that have been fulfilled by the project.

Assessment of BHOOMI by Albert Lobo and Suresh Balakrishnan under the heading "Report Card on BHOOMI" a study conducted by Public Affairs Centre, Bangalore highlights following things.

In all the parameters considered for evaluation, BHOOMI has emerged as favourite in all respects, overwhelming support for BHOOMI can be inferred as most of the objectives of the "Computerization of land records" with respect to accessibility, maintainability, availability and reliability have been met.

Another research study which emphasizes on impact of BHOOMI on RTC services and mutation process by Narayana A. Gatty under the title "Information and Communication Technology for Governance Reforms: The Case of *Bhoomi* Project in India" has come out with following results. While Table 8.2 to Table 8.4 indicate impact of BHOOMI on RTC services Table 8.5 to Table 8.7 show impact of BHOOMI on mutation process.

Table 8.1: Impact of BHOOMI in Land Records Management with Emphasis to Availability and Correctness of Records

Sl. No	*Measuring parameters*	*In BHOOMI*	*Manual System*	*Remarks*
1.	Ease in use of the BHOOMI kiosks	66%	25%	Many of the BHOOMI users (68%) had also made use of the manual system in the past; a majority of users (78%) who had past experience with the manual system found the BHOOMI system more simple.
2.	Complexity in procedure	Most users (79%) of the BHOOMI kiosks got their work done without meeting any official except the counter staff	In manual system 19% of the users had to meet one official, 61% of the users had to meet two to four officials for their work	Legacies of the manual system have not completely faded away. About 18% of BHOOMI users reported that their document was not signed by the appointed village accountant, operating the kiosk. 6% reported that they filled out an application form for issue of an RTC.
3.	Errors in documents	BHOOMI kiosks provided error free documents to 74% users.	In case of the manual system 63% users got error free documents	Among those reporting errors, wrongly spelt names were the most frequent error (81% in case of manual system, and 53% in the BHOOMI system). However, major errors in land details were the issue for 31% of those who reported errors in the manual system, in contrast with 4% in case of BHOOMI users.
4.	Rectification of errors	Almost all users had confidence to complain and sought rectification (93%)	Less than half the number only had confidence to complain and get errors rectified (49%)	Given that errors are not unusual at this stage of development of the BHOOMI system, one need to concentrate on how efficient is the response systems? Half the complainants (58%) got timely response in case of BHOOMI, while such response was reported by only 4% of those using the manual system.
5.	Cost of service (corruption and bribery)	Only 3% of the users of the BHOOMI system reported paying bribes.	Two-thirds of the users of the manual system paid a bribe – 66% of them reported having to do so very often	All users of the BHOOMI facility who wish to receive a hard copy of the RTC are to pay a fee of Rs. 15/- each and receive a receipt for the same. A large segment of users (66%) reported that they did not get (collect?) a receipt for the payment they made.
6.	Cost of service (Hidden cost)	Most BHOOMI users (79%) reported a minimal waiting time in the queue of 10 minutes or less	Only 27% could meet the concerned official in such short time.	The bigger issue is the number of times a citizen had to visit these offices to get the certificate. While most users got the RTC (72%) with one visit to the BHOOMI kiosk, only 5% got it that fast in the manual system.
7.	Staff Behaviour	Most BHOOMI users (85%) rated staff behaviour at the BHOOMI kiosks as 'good'	None of the users of the manual system rated staff behaviour as 'good'.	While technical capacity of the system plays an important role in its success, the approach of people who handle the task is of critical significance too.

Table 8.2: Average Time Required for Obtaining Copies of RTC Before and after the BHOOMI

	Before BHOOMI	*After BHOOMI*	*Time saving*
Southern Karnataka	2.3 days (4.6 visits)	1.07 days (1.07 visits)	1.23
Northern Karnataka	2.7 days (5.4 visits)	1.02 days (1.02 visits)	1.68
Total	2.5 days (5 visits)	1.04 days (1.04 visits)	1.45

Table 8.3: Break-up of Percentage of Farmers according to the Number of Days it took for them to obtain RTCs Before and After BHOOMI

	Percentage of farmers who got their RTC before BHOOMI			*Percentage of farmers who got their RTC after BHOOMI*		
	1 day	*2 days*	*3 days +*	*1 day*	*2 days*	*3 days +*
Southern Karnataka	28.3%	50.3%	21.4%	95%	5%	0
Northern Karnataka	30%	63%	7%	91%	9%	0
Average	29.15%	56.65%	14.2%	93%	7%	0

Table 8.4: Cost of obtaining the Copy of RTC Before and after BHOOMI with the break-up of the components of the Cost

	Before BHOOMI					*After BHOOMI*				
	Cost (in Rs)	*Components (in Rs)*				*Cost (in Rs)*	*Components (in Rs)*			
		Fees	*Bribe*	*Trans-portation cost*	*Opportu-nity cost*		*Fees*	*Bribe*	*Trans-portation cost*	*Opportu-nity cost*
Southern Karnataka	197	03	38	41	115	98	15	00	19	64
Northern Karnataka	207	03	41	42	121	94	15	00	23	56
Average	202	03	39.5	41.5	118	97	15	00	22	60

On an average, farmer saves Rs. 142 per RTC as compared to manual system. Considering that a farmer on an average obtains a copy of the RTC at least thrice a year, the annual saving works out to Rs. 315, which is reasonably good saving.

Table 8.5: Average Time for carrying out Mutation in RTCs before and after BHOOMI

	Before BHOOMI	*After BHOOMI*	*Increase/ Decrease*
Southern Karnataka	29 days	50 days	+ 21 days
Northern Karnataka	27 days	55 days	+ 28 days
Average	**28 days**	**53 days**	**+ 25 days**

One can infer two important factors from the above table, though there is a steep hike in the number of days required for the completion of mutation as compared to pre-BHOOMI days, it also indicates that Statutory Notice period of 30 days as per the Karnataka Land Revenue Act and Rules 1964 is being given to farmers to raise any objections on the transaction. There is a clear indication that statutory notice period was not followed in pre-BHOOMI days thereby curtailing the rights of the farmers. Further, if 30 days is subtracted from 53 days which is average time taken for mutation after BHOOMI, we get 23 days which is five days less than the time taken in pre-BHOOMI days.

Table 8.6: Average Number of Times Farmer Visited Village Accountant in Connection with Mutation

	Before BHOOMI	*After BHOOMI*	*Difference*
Sothern Karnataka	4.5	4.3	–0.2
Northern Karnataka	5.1	4.1	–1.0
Combined	**4.8**	**4.2**	**–0.6**

Table 8.7: Average Cost of Single Transaction of Mutation before and after BHOOMI with the break-up of the components of the cost

	Before BHOOMI					*After BHOOMI*				
	Cost (in Rs)	*Components (in Rs)*				*Cost (in Rs)*	*Components (in Rs)*			
		Fees	*Bribe*	*Transportation cost*	*Opportunity cost*		*Fees*	*Bribe*	*Transportation cost*	*Opportunity cost*
Southern Karnataka	2586	35	2347	40	112	2969	35	2731	77	126
Northern Karnataka	2723	35	2511	38	112	3135	35	2896	94	110
Average	2615	35	2429	39	112	3052	35	2813	85.5	118

There has been significant improvement in case RTC distribution services, same is not true with respect mutation process after implementation of BHOOMI. Though there is a marginal improvement in reducing number of visits by farmers to Village Accountant for mutation, there has been substantial raise in the corruption in mutation process compared to pre-BHOOMI scenario. This may be partly due to making up efforts by field level functionaries with respect to what they lost in RTC distribution and partly due to introduction of new players in mutation process in the form of computer operators. Nailing down to exact reasoning and elaborated discussion on this matter is out of the scope of this case study, however, attempts to improve the system is discussed in the latter part of this section.

In the next set of tables, actual reports from BHOOMI data has been given to show average time taken for mutation, age-wise pendency, ranking reports, report showing category-wise number of requests received for mutations.

Keeping the studies referred above in view, BHOOMI is continuously trying to improve and that has been one of the main strengths of BHOOMI in achieving sustainability. Following are few examples of initiatives in BHOOMI to tackle the questions raised in above.

First In First Out (FIFO) policy in all critical stages of mutation process will ensure that nobody can jump the queue, Revenue Department is giving lot of publicity with respect to FIFO policy thereby citizen understands that when his/her mutation turn comes, nobody can stop it. Further, BHOOMI has restricted the reasons for rejection through a master list. Mutation accepting authorities have to select one of the rejection reasons and reject mutation. These rejected mutations cannot come back to BHOOMI without going on appeal to sub-divisional magistrate, necessary checks has been incorporated in BHOOMI. Further, BHOOMI monitoring cell is constantly reviewing the rejected cases to know whether decision was correct or not. This helps in avoiding false rejection and defeat FIFO. FIFO policy to some extent will help in reducing the corruption in the mutation process as indicated in Table 8.7.

BHOOMI-KAVERI integration has achieved many things which have been explained under stakeholders section under the heading

"Registration Department". There are instances wherein requests for mutation pertaining to registration happened in 1930 coming now. It was left to the discretion of the Taluk office staff and field level functionaries to start the mutation. Though there were procedures to perform checks and balance by revenue audit teams, they were implementation challenges which are common to any paper based system. After BHOOMI-KAVERI integration, whole process has been automated, immediately after registration; XML data sent from registration department is used to initiate the mutation transaction in BHOOMI within few minutes and notice for calling objections if any is generated automatically without any human intervention. What was not happening for years in the manual system will happen in minutes now.

SMS integration in BHOOMI and registration process helps the citizen to know the status of his transaction and brings much required transparency in the government process. This also helps the citizen to avoid bribing officials due to lack of information.

BHOOMI services are contributing significantly for sustainability of 800 rural tele centres started under NEMMADI project, the services from BHOOMI such as RoRs, Mutation extract alone amount to an average of Rs. 1,050 per month per tele centre. This is guaranteed income for tele centre in addition to service charges collected for G2C services of revenue department numbering about another 40 services.

Issues and Challenges faced during Implementation

1. Making computerization of land records as primary activity was a herculean task as this was seen as a secondary activity compared to election, census, disaster, etc. It used to lose importance every time any of the activities mentioned above which occurred and one or the other such activity used to be there always. Project Champion struggles a lot to make it primary activity by making every one right from top bureaucracy and political leadership understands that proper maintenance and updation of land records are important for peaceful society.
2. Digitization of legacy data of 20 million RTCs was a real challenge due to:
   (a) Preparing the records for data entry was phenomenal

task as there were different ways of maintaining records in five different regions of the State. About 10,000 officials were trained in data preparation activity.

(b) Non-availability of good data entry agencies at district level was another challenge. BHOOMI implementation team had to train the data entry agency staff also in carrying data entry legacy records.

(c) Total effort was about 20,000 man-months legacy data entry involving 1200 operators.

3. Verification of digitized records was a real challenge; certification of correctness of data entry is prerequisite for any electronic system to become operational. Making field level functionary complete verification and certification was extraordinary task. There was a need to stop people from doing incremental correction instead of one time complete correction. Different levels of hierarchy right from Village Accountant to Deputy Commissioner were assigned with percentage of records to be verified and certified.
4. Rigidity in redefining rules and procedures was another important challenge, all people in the hierarchy had to be convinced to bring amendments and procedural changes in maintaining legacy records.
5. Lack of technical support at taluk level was the major hurdle in implementation, though district headquarters taluks could be operationalized with the help of District Informatics Officers of National Informatics Centre, same thing could not be done as these officers were stationed at district headquarters and had assigned with many other computerization projects. Revenue Department had to recruit one district consultant per district to assist the department in implementation of the project across the State.
6. Maintaining Hardware, UPS and provide consumable such as secured stationery, holograms, printer cartridges, etc., at taluk level was another important challenge. Revenue department had to come out with a concept called facility management to overcome this. Today a single vendor manages computer, printer, UPS, supply of stationery, etc.
7. Training and capacity building was most essential and critical

activity for successful implementation. Whole of revenue hierarchy involved in BHOOMI activities had to be trained. A continuous training in batches were organised for more than one year to cover all officials initially. Now incremental trainings are being organized as and when new reforms are introduced.

Key Lessons

1. Demonstration of Political will is key to the success of an e-Governance project: BHOOMI was the first e-Governance project that the Chief Minister would mention while listing his achievements*—be it in media interviews, government advertisements or speeches in the legislature and other forums. For example, the budget speech of 2001 when CM himself as Finance Minister of Karnataka gave commitments on timeliness of BHOOMI implementation.
2. *Responsive and Committed Bureaucracy*: Commitment from bureaucracy is a critical factor for implementation of e-Governance projects. Revenue department in Karnataka has shown a lot of interest in implementation of BHOOMI right from the day digitization started. All the reforms have been implemented with co-operation and support for the field level staff.
3. *Continuous and Gradual Improvements*: BHOOMI is a finest example of evolving e-Governance application. Over the years functionalities and facilities in BHOOMI have grown multifold. Started as small vanilla application to manage online mutation in 1999-2000 and further rolled out across the State in 2000-2001, now BHOOMI has a widespread scope in terms of bringing orderliness, accountability and transparency in the mutation management using biometrics for user authentication, FIFO techniques for approvals, invalidating manual records, creating the data centre to store and

---

* Government advertisements to showcase the achievements in newspapers during October 2002, 2003; Chief Minister's speech at 50th National Development Council (NDC) meeting, December 21, 2002, New Delhi.

disseminate RoRs through WEB, making the status of transaction available on WEB. PKI (Public Key Infrastructure) has enabled BHOOMI to make it adhere to IT Act 2000. Village level functionaries like Village Accountants and Revenue Inspectors have been provided with Digital Signature Cards to approve the mutations and sign RoRs as digital documents. Survey activities have been appropriately merged with BHOOMI to achieve much required synchronization of textual and spatial data in the field of land records. Electronic integration of BHOOMI with land acquisition has resulted in updated land records without human intervention. Electronic integration of BHOOMI with Banks and financial institutions has been completed.

4. *Receptive and Proactive Software Development Agency*: National Informatics Centre has played a vital role in success of BHOOMI project. BHOOMI is a good example of evolving e-Governance project and most of successful e-Governance projects have evolved over time. Software agency becomes crucial in the scenarios where softwares evolve over time. National Informatics Centre's software development team has provided domain expertise and crucial continuity for the project which is essential for any e-Governance project to succeed and sustain.
5. *Training and Capacity Building*: Training is one activity which has been given maximum importance throughout implementation of BHOOMI project. Continuous training programmes for Village Accountant to Assistant Commissioners have helped for making BHOOMI sustainable to large extent. A well equipped training centre with state of art facilities is part of BHOOMI Monitoring Cell at State headquarters. Training programmes are being conducted by calling subject experts from different departments, achievers in certain reforms in the area of land records management within revenue department, Officials from Survey department and few Deputy Commissioners who have worked on reforms in the area of land records management.
6. *Focusing on Younger Staff*: Operators-cum-Village

Accountants working in BHOOMI back offices in taluks are being considered as the backbone in implementation of the project. Right from the day one revenue department has focused on younger staff for computerization activities in lower bureaucracy. Initially these young people were picked up from compassionate appointments and later exclusive recruitment was made for these posts (Approximately 1200 posts were newly created) where computer knowledge was made one of essential criteria for appointment. Focusing on younger staff not only ease out the effort of training them keeping their grasping power in view but also helps as they are available for long years to come.

7. *United Command to Achieve Total Land Management*: Real benefit of computerization in the field of land records can be exploited only when all three wings of revenue department namely Mutation management (revenue), Registration and survey department work in co-ordination. Computerization in silos with respect to these departments will be bottleneck in achieving total land records management. All three wings should converge at single vision to make life of citizen hassle-free. NLRMP (National Land Records Modernization Programme) also recommends this type of single command. In Karnataka Principal Secretary (BHOOMI and UPOR) is handling computerization and reforms pertaining to all three wings of revenue department. This unique advantage is helping Karnataka to have more and more electronic integrations.
8. *Reduce Human Discretion Wherever Possible*: Any e-Governance systems can monitor the activities only after transactions are digitally captured on them. Even if e-Governance systems are in place and its interfaces are exchanging data in the form of paper, actual benefit cannot be exploited. When orders in the form of papers are exchanged, individuals working there will only decide when to enter the transaction giving undesirable discretionary power to the officials as to when to start the transaction, even if the target system is electronic in nature. Automatic initiation of transaction soon after receipt of electronic data

will come handy in such scenarios apart from making monitoring easy. BHOOMI is converting all its interfaces to electronic in nature one by one. BHOOMI-KAVERI integration, BHOOMI-BHOOSWADEENA integration, BHOOMI-BANK integration which was discussed earlier are examples of this.

Methodology adopted for Case Study Writing

Methodology: Explanatory and Intrinsic, wherein authors have interest in the case and have first hand information with respect to each and every activity carried out during the project conceptualization phase, design phase, development phase and implementation phase.

Types of Evidence: It is combination of Documentation, Direct Observation and Participant Observation.

References

Albert Lobo and Suresh Balakrishnan (2002). Report card on service of BHOOMI kiosks: An Assessment of benefits by users of computerised land records system in Karnataka.

Bhatnagar, S. and Chawla, R. (2004). Online Delivery of Land Titles to Rural Farmers in Karnataka, India.

Bhatnagar, S. and Chawla, R. (2005). Bhoomi: Online delivery of record of rights, tenancy and cultivation to farmers in Karnataka.

Deepa Narayan (2004). Field Visit to Karnataka, Bangalore, The Bhoomi Programme — Facilitators Report FOR World Bank.

Mearns R. (1999). Access to land in rural India. World Bank Policy Research Working Papers No 2123. World Bank.

Narayana A. Katty (2005). Information and Communication Technology for Governance Reforms: The Case of Bhoomi Project in India.

Sharma, B.V. and Bhat, P.V. (2004). — BHOOMI— A case study.

Chapter – 9

Land Governance: Need to Protect and Use Judicially Public Lands

Nivedita P. Haran

Not only in India, but in every country, public lands are held by a Government department or agency on behalf of the State. Given that land is an inelastic asset, it is essential to utilise public lands in a judicious manner in support of a valid public purpose. Put another way, public lands are held on public trust and Government has to act in the manner of a trustee or custodian in the allocation and utilization of public lands.

Unfortunately, public lands have often been allocated for various types of private uses—ranging from commercial purposes to various forms of private organizations—without either a transparent set of norms, an open tender process, lease rent at fair market value, or consideration of whether the allocation of land in question is in the public interest and serves a valid public purpose. These unsound practices have widespread and unrecognized ramifications on the society and economy as a whole, from large increases in the price of land to the monopolization of public assets for a privileged few. This has been the situation in the past decades leading to the anomalies and the lack of well-thought-out policies that now cries for solutions.

This situation, which also exists to lesser or greater extent in other States throughout India, has been criticized by the Supreme Court most recently in Singh *et al.* *v.* State of Punjab *et al.*, (26 July 2010), in which the Court held that State Governments are duty-bound to take concrete steps to evict illegal occupants on public land and requested all Chief Secretaries to submit regular reports.

This chapter emphasises the urgent need to inventorise and protect all public lands, creation and development of the Kerala State Land Bank as the repository of public lands in Kerala and explains valid public purposes for the allocation of public lands. This chaper continues by identifying the problems that exist in the allocation of public lands and explains solutions to those problems with case studies from recent practices. The chapter concludes by listing the clear advantages of protecting and digitising records pertaining to public lands as a prerequisite to computerisation of all land records.

A. Status of Public Lands

Despite the competencies established *inter alia* under the relevant statutes for the Government to conserve, improve and manage public lands, there had been no systematic assessment of public lands vested in the States even 65 years after independence. As a consequence, significant encroachments had taken place on public lands.

Part of Revenue Department's responsibilities as custodian for public lands is to ensure that such public lands are allocated in support of valid public purposes. These valid public purposes for the use of public lands are set forth below:

1. to establish and maintain offices for Government and other public bodies and agencies;
2. to establish and maintain basic infrastructure facilities, viz., roads, public transformers, water and sewage treatment plants;
3. to create and maintain public 'green spaces,' including the establishment of public parks and to protect the environment;
4. to protect of cultural heritage, particularly historic properties and buildings;
5. to ensure equal access for all people to scarce facilities established on public lands;
6. to provide support to members of Scheduled Castes/ Scheduled Tribes and other landless by allocating land to the 'landless' poor in a sustainable manner;

7. to raise revenue, by collecting fair market value rent from commercial establishments located on public lands; and
8. to maintain sufficient public lands in reserve in order to meet future requirements.

Government action in these areas should be guided in part by relevant Supreme Court decisions. For example, in a specific case, the Supreme Court questioned the propriety of a land allotment to a Trust in the name of a former political party leader and held that discretionary land powers of government are not meant to benefit the affluent and politically well-connected, but rather for benefit of the poor.

Moreover, in Jaspal Singh *v.* Government of Punjab (26 July 2010), the Supreme Court has held that a State Government is duty bound to take concrete steps to evict illegal occupants of public land and that a long duration of illegal occupations or political connections or expenditure in making constructions on the property do not constitute justification for regularizing an illegal position. Regularization would only be appropriate in exceptional cases, such as on behalf of the landless or SC/ST or where a public benefit existed on site. The Court decried the use of "muscle power/money power and in collusion with the officials..." to mis-use public lands. See also, M.L. Builders (P) Ltd. *v*. Radhey Shyam Sahu, 1999 (6) SCC 464 wherein Supreme Court ordered restoration of a park after demolition of shopping complex built at cost of over Rs. 100 crores; Friends Colony Development Committee *v*. State of Orissa, 2004 (8) SCC 733 (even where the law permits compounding unsanctioned constructions, that is only an exception; this principle applies in greater force when encroaching on common land of villagers; orders to allot Gram Sabha land to private persons are illegal).

B. Public Land used for Commercial Purposes without Payment of Fair Market Value Rent

In furtherance of its duties as custodian for public lands, Revenue Department's review of allocation of public lands showed that significant amounts of public lands have been allocated for commercial purposes without either an open tender process or collection of lease rent at a fair market value. Additionally, public lands allocated for non-commercial purposes have been converted to commercial purposes without approval of Government. It is clear that the allocation of public lands for commercial purposes without payment of fair market value in lease rent and open tenders leads to distortions in the market for real estate and creates conditions favourable to corrupt practices.

For example, large extent of land is given on lease to individuals and entities in urban as well as in rural areas. These lands are being used for commercial purposes and yet lease rent recovered is a nominal amount. The entire Marine Drive in the heart of the Ernakulam City is held on the basis of 'ground rent' at a nominal rent. Even though commercial malls and shopping complexes are functioning here, commercial rent is not being paid by the lessees. Indeed, revenue generated from public lands is only nominal in many States despite such significant commercial usage.

While certain private individuals and companies have reaped a financial windfall bonanza from this inappropriate allocation or grabbing of public lands, Government has obtained no financial benefit. On the contrary, precious Government lands have been exploited by private parties for their own benefit leaving Government poorer for it. These lands include quarries, river banks, etc. It is not inaccurate to say that the few remaining parcels of public lands in the heart of Ernakulam city, if auctioned off, can fund the entire budget of the State. In fact, what should have happened is that through leasing of such prime land, Government could have earned huge revenue which could easily meet the social and welfare responsibilities of the Government.

The example of Brunton Boatyards (a hotel) in Ernakulam can be used as a basis for policy in the allocation of public lands for commercial purposes. In such cases, Government should either seek to terminate the lease or allocation of property before time or, upon expiration, take action to recover the property immediately. All such lessees should pay a fair market value in rent with open tenders where applicable.

C. Public Lands remains unutilized after Allocation

Significantly, much public lands allocated for various purposes remain unutilized. In particular, public lands allocated for industrial purposes have remained unutilized, thereby making no contribution to State's economic development. These parcels range from amounts of less than 1 acre to large quantities of hundreds of acres of public lands allocated to corporations.

Part of the reason for increases in recent years in the market value of lands in Kerala—one of the highest amongst the Indian States—is the improper allocation of public lands for commercial purposes without charging fair market value lease rent. The failure to collect a fair market value in lease rent enables companies to exploit political connections to obtain public land cost-free, which drives up the price of land, and inhibits open competition particularly where the land allocated is not used. Moreover, there is a strong nexus between crime and land, where most major criminal cases have a land aspect and criminal activity is funded through undue profits made from the uncontrolled land price hike.

This problem also leads to much broader negative effects. In part due to this situation, the purchase of a plot of land in urban areas for building a house is beyond the means of the common man with resulting pressure placed on rural areas and conversion of wetland and paddy land with negative environmental effects.

It is the duty of Revenue Department as custodian for public lands to resume public lands that are not utilized within a given time period. Such conditions have to be incorporated into all leases to enable prompt action to recover possession. Reallocation of such lands will occur only upon payment of fair market value commercial rent and open tenders where applicable. Effective implementation of this policy will increase the supply of land, promote economic development and help stabilize prices in the property market.

D. Public Lands allocated to Institutions claiming a Charitable or other Public Purposes but Actually used for Commercial Purposes

A large extent of public land has been taken on lease by institutions claiming a public or charitable purpose or relationship to sports and culture are in fact being utilised for hard-core commercial purpose, viz., shops and shopping complexes and malls, private offices, etc.

In many cases the land was taken on *kuthakapattom* or lease before independence when public lands were in plenty, there was limited public and civic services and above all no democratic welfare State. Additionally, public land is allocated to Government-aided private educational institutions that charge significant capitation fees to students. These practices promote corruption as Government is encouraged not to monitor whether public lands are used in the manner set out in the lease agreement or other allocation documents. In particular, using public land for commercial purposes when the purposes set out in the lease agreement with Government were charitable in nature may constitute fraud.

For example, Pathrakulam, a traditional pond in the heart of the capital city Thiruvananthapuram has been given on lease at nominal rent to a Trust run and controlled by the family of an ex-Chief Secretary. This property after being mercilessly filled up is being leased on commercial terms for exhibitions and other commercial purposes. The Irrigation Department has reported that one of the reasons for the monsoon flooding in the city is the filling up of Pathrakulam.

In the case of capitation fees, schools that levy such charges often themselves are recipients of significant State aid. Yet, they do not make provision for scholarships to benefit the disadvantaged and generate significant revenues through capitation fees. Moreover, Government is investigating many of these schools for inflating their enrolment figures in order to increase the State subsidy allocated to schools as well as to maintain control of the land. In cases where enrolment is inflated, these private schools should be consolidated to make more efficient usage of scarce public land and to reduce the drain on the budget from State subsidies to support non-existent students.

Government has a duty to investigate the allocation of private lands for allegedly charitable, educational and cultural purposes to ensure that fraud and unjust enrichment do not occur through the illegal conversion of these public lands to commercial use.

E. Private Clubs on Public Land: Monopolization of Scarce Facilities and Improper Commercial Usage

A number of private clubs operate on public land where lease rent is overdue for years without action being taken by Government.

Other problems exist as well. In some cases, only nominal rent is paid for valuable public land that is monopolized by members of a private club. Additionally, many of these establishments earn revenue by hiring out their premises for weddings and other celebrations, and are, therefore, in use as commercial premises that compete with legitimate businesses. Moreover, in some cases public monies budgeted for tourism and other uses have been used to pay for renovations and reconstruction of premises at these private clubs. These expenditures have no rational relationship to any public purpose and may indicate corrupt practices at work.

In the much-reported case of the Trivandrum Golf Course, the Supreme Court upheld the State Government stand to resume the land. The land has now been taken over by Government with plans in place to lease the course and other facilities for management upon open tender and payment of lease rent based upon fair market value. All those interested will now be able to play golf and use the facilities that will now be open to all members of the public.

The "Golf Club" case is particularly important in view of the history and uses of the land. During British times, the colonizers established a golf course in Trivandrum with land obtained from the then rulers and used the property as a private sporting club from which common Indians were excluded. After Independence, the golf course became public land owned by the State. As the only golf course in Trivandrum, the public land should have become a public golf course open to all members of the general public. Instead, the golf course was allocated to a self-appointed Indian elite who monopolized the premises in the form of a private club.

In these cases, it is essential to distinguish between: (a) the public land and the facilities established on public land, and (b) the members of a private club. Members of such clubs may continue their good fellowship and play golf at the golf course that is open to the general public, but they should not be able to exclude the general public from the facilities or profit from commercial activities at facilities established on public land.

If a group of individuals wish to establish a private club for playing golf or tennis, then they are free to pool their private monies and purchase land on the open market to establish such a private sporting club. These private individuals, however, should not be able to monopolize the use of facilities for sporting activities established on public lands, usually with public money. This constitutes a corrupt practice with a huge social effect as these sporting facilities are limited and should be open to the public rather than being monopolized by a privileged self-appointed elite.

There is also a negative impact on the development of Indian sport, as these facilities often constitute the only facilities available for playing these sports and are not open to the general public. The practice of denying the general public access to scarce sporting facilities in favour of reservation of those facilities located on public land in prime locations for the benefit of a privileged self-appointed elite speak of corruption.

This principle of distinguishing between the facilities established on public land, often with public money, and the members of a private club should be applied as well to all establishments.

These facilities should be taken over by Government and put out to tender with special provisions to ensure lessons for youth groups and the disadvantaged as well as for school-based sporting associations.

F. Alienation of Public Land allocated to the Landless

Under the Kerala Land Assignment Act public land may be assigned to landless poor. The Kerala Land Reforms Act had the basic intention to take over excess land from 'big landholders' and assign such surplus land to the landless labourers. Based on the Land Assignment Act, Government also had a policy to distribute *poromboku* land (public lands) to the landless for the purpose of self-housing, agriculture and in some cases for beneficial enjoyment, subject to the condition that these lands are included in the list of assignable land or '*tharissu bhoomi*'. In order to be eligible for such assignment the annual income of that family should be within Rs. 30,000, which was raised to Rs. 75,000 in 2010.

Yet, this policy has been undermined by alienation of land allocation to the landless often to developers in sales below market value. Part of the problem is the multiplicity of standards regarding alienability of land allocated to the landless. Under the Land Assignment Act, the period was 9 years. In the case of surplus lands assigned under '93 Special Rules, the period was 3 years. In the case of surplus lands assigned there was no time limit. The confusion that emerged due to this was taken advantage of by the assignees who were keen to alienate. The officials were also happy and allowed the assignees to take benefit of whichever category they prefer. No clear policy was laid down to ensure that the land assigned was actually utilised by the assignee for his personal use and not for commercial profit.

In Munnar, for example, regularization of the use of public (forest) land under the 1993 rules was abused through the use of *benami* to grab forest land in large extents for the running of resorts and hotels. Regarding land assigned to the landless either by way of surplus land or public lands, a quick survey conducted by the State Land Bank has revealed that in more than 60 per cent of the cases land that is assigned to the landless is alienated within a time period of one year (which is illegal). Only in less than 20 per cent of the cases it was found that the original assignee was actually occupying and living on the land after 10 years or more.

The issue of lands assigned to tribals but taken away from them and resorts set up on those lands, is especially serious. Even though such alienation is illegal this has been happening very often. It makes a mockery of a noble policy to make all tribals landed in the next 2 years. While the tribal receives Government land with one hand, he sells it off to a hotelier or a commercial entity with the other. At this rate tribals will forever remain landless. Hence, Government should plan land assignment to tribals in such a way that their houses are clustered in one area, with required infrastructure but no permission to alienate before a minimum of 25 years.

G. Conversion of Possessory Interest in Public Land into Freehold

In a number of cases, public land has been converted into freehold interests for private parties in an improper manner.

In one case, a major resort on Kovalam Beach owned by the State and on land obtained through land acquisition, acting through the Kerala Tourism Development Corporation was given on lease to the India Tourism Development Corporation. That leasehold interest was subsequently converted into a freehold interest on behalf of a private party and is now trading as a hotel and resorts. Yet, a leaseholder cannot transfer an interest in real property that the leaseholder does not possess. Thus, no legal basis existed for the ITDC to sell a freehold interest in the subject real property. This case is particularly important because the property in question includes heritage buildings for which general public access is now denied by the private party. Moreover, the private party in question has blocked access to the public road along Kovalam Beach, forcing local residents into a detour of several kilometres as well as effectively privatizing the beach itself.

In another case, a 40 acres plot in Thrissur was allocated to Crompton & Greaves on lease. Surprisingly, a proposal was made by the company with a supporting recommendation from the Industry Department to allow the leaseholder to sell the freehold. No basis exists to allow a leaseholder to sell a freehold interest in Government property and then pocket the money. If it is determined that the sale of public land is in the public interest, then such sale should only be done by open tender with a reserve price auction to ensure that Government obtains the benefit from the same of such public land.

All cases of conversion of public land into private freehold interests should be investigated by Government.

H. Public Land and Assured Infrastructure

Misallocation of public land also leads to discriminatory effects in the development of public infrastructure. In Kerala, a State with high population density and virtually one urban continuum along its 590 km coastline, less than 30 per cent of households in cities and towns have piped water supply and less than 20 per cent are connected to a sewerage system. Hotels and resorts, for example in Kumarakom, obtain a tremendous benefit in the form of assured infrastructure. Dedicated transformers provide permanent electricity supplies whilst the local inhabitants face load-shedding.

This discriminatory practice should cease. Tourism will improve if infrastructure for the entire community improves, not just for tourist enclaves.

I. Improper Assignment of Road and River Poramboke

In a number of cases, road and river poramboke have been allocated to private parties. This type of allocation of poramboke land leads to disastrous effects on economic development and environmental protection in Kerala.

With respect to roads, it becomes difficult to widen roadways without significant land acquisition procedures. In the case of the national highway, planners at the Centre stopped construction in Kerala due to difficulties in land acquisition to widen the roadway.

With respect to rivers, the result is encroachments on backwaters and *kayals*. It is necessary to survey, demarcate and remove all encroachments in order to prevent release of effluents into backwaters which is a health risk for people living along the bank.

This problem also has broad negative effects, such as on the food supply. Kerala is a food deficit State in areas of cereals, fruits and vegetables as well as meat and poultry. At the same time, Kerala's traditional paddy lands are getting filled up willy-nilly. A number of scientific studies have forecast that with the large-scale filling up of paddy lands into residential plots, private educational institutions and commercial use, this is having an adverse impact on the State's environment, causing watertable to deplete or saline intrusions sometimes upto 5 kilometres within the coastline.

J. Failure of Government to enforce existing Orders leading to Encroachments

In a number of cases—some of which date back decades—Government has failed to take the necessary actions to complete the resumption of public lands. These cases particularly include 'big landowners' that are the subject of longstanding legislation on land reform. By allowing these cases to lie fallow and unattended, Government has distorted the market for land because these large tracts are not available as part of the State Land Bank. All such cases need to be investigated and the land resumed by Government forthwith.

K. The Kerala State Land Bank

The Kerala State Land Bank has been set up as a mechanism to ensure that public lands are allocated and utilised only for valid purposes.

In order to create professional and transparent governance of the lands in the public domain, in 2008-2009, in a landmark move and the first of its kind in the country, Kerala established the Kerala State Land Bank under which the Department of Revenue would act as the custodian of public lands and hold responsibilities as trustee for the inventory, surveillance and protection of public lands, as well as for revenue generation.

The role of Kerala State Land Bank in three-fold:

1. Inventorise all public lands;
2. Demarcate and protect these lands; and
3. Use public lands for public purpose and for generating revenue without alienating them.

The Kerala State Land Bank set up in 2008 is now fully functional. Details of all public lands are posted on a website. All lands are demarcated and special *jhandas* (stones) are being placed. A Special Protection Force has been set up to directly inspect each parcel. Given the progress, other departments and agencies have also started adopting similar mechanism.

The State Government is now examining the proposal to set up a Land Bank Society for the purpose of:

1. maintaining and updating land records;
2. completing geo-referenced re-survey; and
3. carrying out all activities of Land Bank.

Government of India policies should also be conducive to protection of public lands. Ministries, PSUs and other agencies should ask for land for projects with adequate justification of the extent and a proper DPR. Once land is assigned by the State Government, the land should be used within a reasonable period. Tendency to play off one State against another only creates unhealthy competition. Public lands are the source of encouraging crony capitalism, inducing corrupt practices in elections and in public life.

Further, privatisation of PSUs should not include sale of public lands, especially lands acquired under the Land Acquisition Act as the same is not legally permissible without the State assent. Often, land is the only valuable asset available with the PSU. In Kerala large tracts of lands with PSUs is the only big land resource available for important and prestigious projects. States should be allowed to resume these unutilised lands to use them for various essential public purposes. In a landmark case, the High Court of Kerala has ordered that when land is acquired and assigned to a PSU for a specific use, its use cannot be altered. In other words, industrial use cannot be changed to residential or commercial.

The advantages that emerge from protecting public lands are too crucial to ignore even though there are very strong vested interest against it:

- By inventorising and consolidating public lands, these orphaned lands would find a protector.
- Large extents of public lands held illegally would thereby come to light and would be a method to tackle graft at source.
- Gram Sabha lands, grazing lands, etc., and even pockets in urban centres can be made part of the town/city Master Plans to enable an urban centre to grow with all infrastructural back-up.
- When consolidated data of all public lands is available, the State can provide these lands instantly, especially for time-bound projects.
- Finally, any computerisation of land records has to begin with public lands.
- There should be a strong input on 'management of public lands' to IAS probationers.

Ministry of R&D would do well to issue guidelines to States advising them to set up Land Banks to retrieve whatever little remains of public lands.

Chapter – 10

Innovative Initiatives in Governance of Land in Kerala

Sajith Babu

Land is becoming a dearth commodity in a populous State like Kerala. The pressure on land would become hundred fold higher in the near future which warrants better land governance systems. Through the project Kerala State Land Bank, the data on Government land have been inventorised and is made transparent and available by creating the landmark website www.kslb.kerala.gov.in for ensuring a participatory governance of Government lands. For the first time in India, only this initiative could bring in 100 per cent transparency in the governance of Government lands. Considering all sorts of land related frauds and civil disputes in private lands and also criminalization activities like dubious land purchases in strategic locations of Kerala by the unwanted elements, the land governance systems like ReLIS [Revenue Land Information System] has become the need of the hour.

Access to land and its security have significant impact on growth and development of the communities, irrespective of diverse environments. In most of the societies, land has been a topic of conflict like that of dearth situations in Kerala. Such conflicts generally end up either in criminal or civil court which subsequently remains as a perennial cause of conflict for the future generations as well. The land related conflicts mostly arise because of the discrepancy between the demand and supply of land between the communities and economies of the world. Even within the countries the disparity is quite obvious among the States as well.

The new world economics hallmarked by commercialization, industrialization, urbanization, etc., puts extra pressure on available land resources, no matter the land in question is under private or public ownership. This results in encroachment and misappropriation of public lands and unprecedented disputes over the ownership of private lands. The figures show that the occurrence of such incidents has been on the rise especially after the 1990's in the State of Kerala. The Governance of Land including inventorisation, protection of public lands, registration coupled with online—real time transfer of registry, computerized, retrievable common land data base, etc., are going to be the key elements in the management of total governance in a State, as land is going to be one of the most critical factors of production, source of investment and a valuable reserve for the future development. The macro and micro level reflections of this phenomenon and associated conflicts are already visible in the world politics and even at the micro social systems like families. Without proper land Governance systems these State of affairs cannot be tackled in an appreciable manner.

Governance of land being one of the early evolved systems, the transformation of the same usually pose lot of challenges in various fronts including multiple legal challenges as it involves plenty of stakeholders with conflicting interests and regulations. The technological advancements of recent times should act as the 'tools of change' and the beneficiary stakeholders should as 'change agents' towards bringing in transformation in the land governance systems. Transformation of the old systems cannot be implemented overnight and it requires perfect planning and execution of the 'change'.

In general, from the administrative point of view, **Land** means "the solid ground portion on earth primarily classified based on tenure as Public land and Private land. The private land refers to portion of land on which there is absolute freehold right and is conveyable by a deed and public land means the unappropriated land belonging to Government, used to encourage economic development of public at large, reserved for public use and for the settlement of the landless". This definition is in conformity with the various Acts and Rules prevailing in India and in most of States including Kerala. The **Land Governance** "concerns the Rules, processes and structures through which decisions are made about access to land and its use, the manner in which the decisions are implemented and enforced, the way that competing interests in land are managed" [FAO, 2009]. Though, the land governance systems have been highly institutionalized since the beginning of the twentieth century in Kerala, the lack of discipline had resulted in unprecedented issues like tampering of land records, frequent incidents of encroachments on government lands, faulty land registrations, etc.

As per the Rules of land and the draft land policy of State of Kerala good governance of land include;

- The access to information on land should be equitable in nature. The degree of information on land is the key to good land governance for the government and the people. The transparency in land deals both by the Government and the private persons can only be ensured through making the information on land equitably available to all.
- *Assured Title and Security of Tenure*: The Land assignment Act 1960 and the Land Reforms Act 1963 could address these issues in a substantial way in Kerala. But the implementation of Transfer of Registry Rules 1966 has to go a long way towards making the tenures legally secure for the communities and the Land Titling Bill 2010 needs to be implemented to redress the related issues in this front.
- *Accountability on protection of public resources based on Doctrine of Public Trust*: The public land resources are to be conserved, protected and reserved for the future generations towards sustainability of the generations to come. The fixing of accountability on protection of public lands has been further streamlined through the enactment of Kerala Land Conservancy [Amended] Act 2009 which imposes serious penal provisions against the offenders including the irresponsible officials.
- Effective and efficient administration of land to cater the G2G, G2C and C2C services: In a democratic country like India the services related to land given by the Government, should be based on an error free inventory of private lands and should satisfy the immediate needs of the citizens within the shortest period of time. This inventory should not override the inventory of public lands which are meant for the public

at large. The situation demands a perfect textual and spatial database generation with the help of ICTs to cater to the immediate needs of the citizens.

Towards ensuring good Land Governance in the State two initiatives have been introduced in the year 2007-08 by Government of Kerala namely the Kerala State Land Bank Project and ReLIS [Revenue Land Information System]. A version 2.0 of Land Record Computerization Programme now being pursued under the NLRMP [National Land Record Computerisation Programme] under the auspices of State IT cell-Revenue.

I. Kerala State Land Bank Project

As per the original objectives of the project it will cater the following services to Government in the land governance sector:

(a) Inventorisation of Public Lands: Under this objective all the public lands are being surveyed and details are hosted on the website **Error! Hyperlink reference not valid.** project was started in the year 2008-09 and as on today, textual database of 25752 land parcels, spatial data base of 5269 land parcels where resurvey has been completed and field survey using WGS-84 datum has been completed for 278 land parcels in villages which were not resurveyed so far. The photographs of 5269 land parcels have also been uploaded in the site.

(b) Income Generation from Public Land: The KSLB will be the sole agency in the State which will act as the custodian of leasable/ licensable Government lands. The entire management of lease cases will be carried out through the **Lease Alert System [LAS]** which is at the final stage of implementation. This would help the Government in maximizing the returns through timely collection of leases without any lapse.

Besides, the land in the commercially important sites will be put to auction for inviting **commercial projects from Public Sector agencies on BOT basis** for 25-30 years. The building plans approved by KSLB, will alone be allowed at such sites mainly towards infrastructure development in the State.

Towards protecting large extents Government lands in rural areas from encroachment, with a better public attention, **Tree Banks** will be established this year with trees like Mahagoni, Teak, Rosewood and

Sandal wood. Considering multipurpose objectives of environmental protection, better returns from Government lands, sustainable income for the Government, protection from encroachment and a measure to meet future wood requirement of the society, this sub-programme will be introduced this year under the annual activity of KSLB.

During the imminent plan period, KSLB is planning to enter a new sector wherein the KSLB will act as **Land Transaction Consultants** in the Government sector which will buy land from agencies like Commercial Banks who are finding it difficult to manage the attached lands during loan recovery process. In addition, the KSLB will buy land from open market to reduce forced/distress sale by the land owners in the State. The KSLB after the inventorisation process is going to prepare a comprehensive list of saleable [Units of 5 cents each] lands. Such Government lands which will be available for the low/medium/high income group of landless people with a vision on development of rural areas. The exhaustive list of such lands will be on display on the website www.kslb.kerala.gov.in and in media also, that the people can apply for such notified land. This type of allotment, based on the quota of allotment will be as per Kerala Land Assignment Rules 1964 which curbs the misuse of such allotted land. The quota for each category will be fixed based on economic and social backwardness of the applicant and after being cleared by the respective Taluk Land Assignment Committees [TLAC]. However, the proposal is still under the consideration of the Government.

(c) Surveillance and Protection: The KSLB with its own machinery and with the help of local revenue officials has been undertaking the surveillance activities of the Government lands. Towards giving a new impetus to this activity, creation of a Public Land Protection Force (**PLPF**) is in the anvil and is at the final stage of implementation. The personnel from PLPF and the Village Officers will sign the **Beat Registers** to be kept near on monthly intervals so as to assure two inspections to a site in a month.

A 24×7 Toll Free Number **1800 425 5255** has already been established towards reporting of encroachments of public lands by the general public and also to inform about the Government lands which are left out from inclusion in the website. It is proposed to make the toll free number an Interactive Voice Response System [IVRS] during the annual plan period 2012-13. Protection of

Government lands is also being done by erecting Boards and by erecting live/non-living fences around the sites.

II. ReLIS [Revenue Land Information System] Land Record Computerization Programme, Version 2.0

The land record computerization activity in Kerala was started in the year 1988-89 along with other States under a programme called SRA and ULR [Strengthening of Revenue Administration and Up-dating of Land Records] with a major thrust on developing the core IT infrastructure. The programme was revisited by the Central Government from 1990-91 and the name of the programme had been changed to CLR [Computerization of Land Records]. The same programme was continuing till 2009-10, when Government of India introduced another programme NLRMP [National Land Record Modernization Programme] to replace all the land record service related programmes in the State. The basic objective of issuing RoR [Record of Right] to all the landholders in the State remained the same by ensuring online transfer of registry [mutation of records] pursuant to land transactions.

As there were lot of issues with the processes involved and to bring in GPR [Government Process Re-engineering] in the system a State IT Cell under Revenue Department was created in November 2008. Soon after the formation of the State IT Cell-Revenue a system analysis study was carried out to differentiate between the VAs *vs.* NVAs [Value Activities *vs.* No Value Activities] involved in the process. Based on the study a planned Government Process Re-engineering [**GPR**] was carried out in the following areas:

Naming and Owning the Project: There was no project name as such for the land record computerization programme. As the IT Cell team has been so much confident of meeting the objectives, a new project name is given to the process of land record computerization programme as ReLIS [Revenue Land Information System]. It is expected that the programme would be one of the path breaking projects ever witnessed especially in a land dearth State like Kerala where land related issues are the order of the day.

Software: The software which had been in use for the land record computerization activity was the BHUREKHA which was developed by NIC. It had many inherent issues including the issues of future financial liability for the State as the same was on proprietary platform. It was not having online applicability and was least customized to fit into the domain requirements for the future application. The new software ReLIS developed through the joint effort of IT Cells and NIC, Palakkad has 100 per cent domain requirements with provision for future add-ons. The most important feature of the software is that it is on FOSS [Free and Open Source Software] and the data already created in BHUREKHA could be easily migrated without any loss.

Hardware: For the first time in the history of Revenue Department a State-wide AMC for all the hardware has been floated since 2010-11 and a system inventory software HaMoS has been developed with the help of NIC, Idukki for the hardware management and monitoring. The SLA [Service Level Agreement] with the AMC [Annual Maintenance Contract] vendor has been rated one of the model SLAs in the sector making the vendor more accountable.

Human Resource Development: The trainings for the officials concerned has been made need based and the *On The Job in House Training Programme* inculcated in the offices proved to be very effective on cost and efficacy-wise. New training module has been initiated on e-District and ReLIS by the ILDM [Institute of Land and Disaster Management] the training centre for Department of Revenue, Kerala.

Management: The centralized control and decentralized monitoring have been successfully tried through the establishment of District IT Cells in each district. The District IT Cells are being assisted by Rapid Response Teams [RRTs] of computer experts identified from within the department, the database of whom was collected in advance and safely kept by the State IT Cells to use at the times of emergency.

Legal Changes: In order to give legal support to the activities, the legal amendments have been made in Transfer of Registry Rules, 1966 besides issuing number of circulars and Government orders towards making a congenial legal environment for the smooth implementation of the much awaited online transfer of registry.

After so much of internal and external brawl, the project launched on 25.10.2011 in 16 pilot villages under the jurisdiction of Alathoor and Vadakkanchery SROs of Palakkad district. As on today 403 numbers of digitally signed Bhoorekhas [RoRs] could be issued online through the Akshaya Common Service Centres [CSCs] besides doing 32 fully automated online Transfer of Registries without any physical movement of connected papers or the citizens on FIFO [First In First Out] basis. As per the plan, this process transformation of Transfer of Registry will be rolled out in 207 more villages in the districts of Pathanamthitta, Kottayam and Alappuzha in the next three months, and altogether by the end of the year, total 557 villages [in 9 districts] out of the total 1634 villages in Kerala will be practicing Online Transfer of Registry and issuance of Digitally signed Bhoorekha [RoR] through CSCs.

The RELIS [Revenue Land Information System] is a multiuser application and this is going to create history in India, wherein the transfer of registry would be made online assuring the following benefits:

1. This ensures transparency in all land transactions.
2. This will check transaction of Government lands.
3. The landholder can get Bhoorekha [RoR] from any of the Akshaya centres in Kerala.
4. It curbs the entire corruption associated with land transactions.
5. Government will obtain more income as it will be linked to fair value of land.
6. The suspicious land transactions in the sensitive areas would be curbed.
7. The Land Reforms Act can be implemented in a more meaningful manner.
8. The transfer of registry is assured promptly and swiftly in FIFO principles.
9. It is a great leap towards assured Citizen Services.
10. Better credit flow under assured land records.

Conclusion

The twenty first century India requires just not good governance but smart government to cater to the aspirations of the technologically advancing generations and communities. Land being the only resource which does not have a horizontal growth, the governance of the same requires much more serious attention on part of the governments in power. The initiatives like Kerala State Land Bank is not only acting as a permanent data bank on Government land but also a big repository of information for the government to plan development projects without waiting for reports from grass root level functionaries. The initiative in addition ensures the transparency in all government land transactions and thus enhances the people's vigil over the governance of public lands. The ReLIS programme is the first of its kind in India in which the entire transfer of registry has been transformed to bring in transparency, speed and FIFO principle through the introduction of ICTs.

Chapter – 11

Problems and Experiences of Autonomous Council with Special Reference to Karbialong and Dimahasao District of Assam

Kishor Mohan Bhattacharyya

Introduction

The autonomous division is known as that while the Central Government has given varying degrees of autonomy within the State legislature. The establishment and function of most of these autonomous councils are based on Sixth Schedule of the Constitution of India. The autonomous council in Assam are as follows: Bodoland Territorial Council, Karbialong Autonomous Council, Dima Hasao Autonomous Council.

Objectives

Our present study is an attempt to—

1. To analyse the problems regarding lands faced by the people lived in autonomous council.
2. Various loopholes in administration system for offering the grant and facilities to the people.
3. And the way how to access their rights provide by the Constitution of India and other facilities.

Analysis

The Constitution of India had given almost unlimited powers to the State Government in respect of local-self governing institutions. This power has been used to centralize power at almost all levels. autonomy is rarely seen in the context of justice further it result to some extent a challenge of assimilation of different tribes mainstream of Indian society. Hence, it is rare to find an instance where autonomy has sought to work on the principle of restitution, by acknowledging that an injustice has been committed. Autonomy—as framed within a statist discourse—does not address the issue of custodianship of resources in a comprehensive manner. The provisions of Sixth Schedule, Article 371(A) and even the recent Panchayati Raj Bill make an effort to ensure the devotion of powers of administration and governance to the grass roots in autonomous council.

Hidayatullah (1979) called the Fifth and Sixth Schedules "Constitutions within a Constitution". When the province of Assam was instituted in 1874, some parts were administered under Scheduled District Act. These were the Backward Tracts which included Lushai Hills (Mizoram), Naga Hills (Nagaland), Garo Hills, the North Cachar Hills, the British portion of Khasi and Jaintia Hills, and Eastern Frontier Tracts of Lakhimpur, Balipara and Sadhyia. The Backward Tracts were part of larger entity called Excluded Areas. After India's Independence the excluded areas were constituted as Autonomous District and Autonomous regions.

The Sixth Schedule divided Tribal Areas in Assam into Part A and Part B. Part A included United Khasi and Jaintia Hills, Garo Hills, Lushai Hills, Naga Hills, North Cachar Hills and Mikir Hills. Part B included North Eastern Frontier Tract including Balipara Frontier Tract, Tirap frontier Tract, Abor Hills and Misim Hills and the Naga Tribal Areas. But the Tribal Areas after 1972 reorganization of North-East India as described as Part I, Part II and Part III. Part I relates to the territories of Assam where as Part II concerns the State of Meghalaya and Part III the State of Mizoram. Part I contains the North Cachar Hills District and the Mikir Hills.

Karbi Anglong originally the territory of the present district was a part of Dimasa Kingdom till the end of Sengya Tularam Hasnu in AD 1854. After his death the territory was governed by British and it was declared a 'Scheduled District'. It was constituted in 1874. On November 17, 1951 a new district, the United Mikir Hills and North Cachar Hills district was created by combining some parts of the districts of Sibsagar, Nagaon, Cachar and United Khasi and Jayantia Hills district of present day Meghalaya State. On 2 February 1970, this district was divided into two separate districts Mikir Hills district and North Cachar Hills district. In 1976, Mikir Hills district was declared as **Karbi Anglong** district. In 2006, the Government of India named Karbi Anglong one of the country's 250 most backward districts. It is one of the eleven districts in Assam currently receiving funds from the Backward Regions Grant Fund Programme.

The main district was divided into three sub-divisions. These are, Diphu, Hamren and Bokajan. Diphu sub-division is further divided into two revenue circles: Diphu and Phuloni while Hamren sub-division contains one revenue circle: Dankamukam Revenue circle; and Bokajan sub-division consists of one revenue circle: Silonijan. It has four Vidhan Sabha constituencies. These are Bokajan, Howraghat, Diphu and Vothatlangso. There is 2633 number of villages in this district. The four Assam Legislative Assembly constituencies in this district are designated for scheduled tribes. They are all in the constituency. According to the 2011 Census Population of this District is 965,280. The population growth rate over the decade 2001-2011 was 18.69 per cent. The sex ratio is 956 females for every 1000 males and literacy rate is 73.52 per cent. A number of indigenous peoples lived in this district. The Karbis are the most prominent amongst them. Other indigenous peoples are Rengma Nagas, the Dimasas, the Bodos, the Kukis, the Garos, the Tiwas, the Khasis, Hamars, the Mizos and the Chakmas. **Karbi** language is one of the prominent languages of this district. Other languages like Bodo, Kuki, Assamese and Aiton. Approximately 1500 speakers is closely related to Shan and written in the Burmese script. A hilly district of Assam Karbi Anglong is a famous place for its natural beauties. Multi-coloured cultures and traditions made up the district as unique. **Diphu** is very busy town. The town is located on a hill. **Silbheta** is a beautiful picnic spot located at 37 km away from Diphu. The place surrounded by rain forests. **Akashiganga** 65 km away from Diphu, a divine stream of water flows down from the high hill in this plot. **Amreng:** Amreng is a very attractive place in Hamren. Here a blue water stream falls down from a huge rock. **Garampani:** Garampani is a sanctuary of Nambor Reserve Forest. There is hot water spring.

Dima Hasao district (DHD): "Dima Hasao" means "Dimasa Hills" in the Dimasa language, before rcognised as a Dima Hasao district it was known as **North Cachar Hills district**. According to 2011 Census it is the least populous district of Assam. Dima Hasao district was a part of the Dimasa Kachari kingdom before 1832. The present North Cachar Hills district was included in the old Cachar district while Asalu was only police outpost. In 1880, this portion was constituted into a sub-division with Headquarters at Gunjung under Cachar district.

In 1895, the headquarters was shifted to Haflong. In 1951, North Cachar Hills was specified under paragraph 20 of the Sixth Schedule of the Constitution of India and excluded to be a part of Cachar district. This part along with Mikir Hills created a new civil district which is "United district of North Cachar and Mikir Hills" with effect from 17 November 1951. According to the provision of Sixth Schedule, two different councils were constituted later on, viz., North Cachar Hills District Council and Mikir Hills District Council within the geographical boundary of that district. The Dima Hasao District Council was opened on 19 April 1952. In 2 February 1970, Government declared an independent administrative district, viz., North Cachar Hills District with the geographical boundary of autonomous North Cachar Hills district council. At present this autonomous council act as administrative controller over almost all departments of the district except law and order, Administration and Treasury Department. The district headquarters are located at Haflong. Dima Hasao district occupies an area of 4,888 square kilometres. It is the third largest district of Assam after Karbi Anglong and Sonitpur district. Dima Hasao district is an Autonomous District (Lok Sabha Constituency).

The Dima Hasao District is administered by Dima Hasao Autonomous Council (DHAC). The Members of the Autonomous Council (MAC) are elected by people of Dima Hasao. The Dimasa of Dima Hasao District celebrates various agricultural festivals in different ways and at different times. The Dimasa festival can be divided into community festivals and local festivals. The local festivals are performed by each village separately, and participation is restricted to the people of the village concerned. The community festivals observed in this district are like Misengba and Bishu. And local festivals are Khorongfang Gerba, Maisalai Gerba, Rajini Gerba, Horni Gerba, etc. In 2006, the Government of India named Dima Hasao one of the country's 250 most backward districts. It is one of the eleven districts in Assam currently receiving funds from the Backward Regions Grant Fund Programme. According to the 2011 Census, the population of Dima Hasao district was 213,529 which is equal to the nation of Samoa. The population density of this district is 44 inhabitants per square kilometre. Its population growth rate over the decade 2001-2011 was 13.53 per cent. The sex ratio is 931 females for every 1000 males and a literacy rate 78.99 per cent. The major community residing in this district are Dimasas, Kukis, Zeme Nagas. Other ethnic groups are like Biates, Hrangkhawls, Vaipheis who are also Kuki, Karbis, Khasipnars and Khelmas. Dima Hasao is the highly Populated by Dimasa tribe. The Dimasas constitute 43 per cent of total population. Whereas Non-Dimasa tribes and Non-Tribals like Bengali, Jeme Naga, Nepalese, Hmar, Kuki all together constitute 53 per cent of total population. The most prominent languages of this district are the Dimasa, Zeme, Hrangkhol, Kuki, Khasi, Biate, Hmar, Karbi, Khelma, and few languages of Indo-Aryan like Haflong Hindi, Assamese and Bengali. The district is globally known for a place named Jatinga, a village, which is famous for the occurrence of birds "committing suicide". According to ornithologist Dr. Anwaruddin Choudhury, the birds, mostly juveniles and local migrants, are disturbed by high velocity winds at their roost. When the disturbed birds fly towards lights as refuge they are hit with bamboo poles and killed or injured.

Problems and experiences regarding land reformation and administration in the autonomous region are as follows:

Shifting Cultivation

Shifting cultivation has been used to justify the forced resettlement of indigenous peoples in country like Bangladesh, Burma, India, Indonesia, etc. The problem raised in fuel land conflicts, increase poverty and open up forests to competing interests. The national policies towards indigenous land rights and those towards shifting cultivation and forest management should be addressed together for their protection.

Gender Issue in Asia

Land ownership in many parts of Asia have recognized women as having equal rights to land as men and in some areas matrilineal inheritance is common. Processes such as the introduction of individualized land holding in indigenous areas, forced resettlement, compensation, registration of heads of households for taxation or benefit-sharing purposes, and the provision of jobs in extractive industries have all tended to select males over females. It seems to us that erosion of indigenous women's rights and resulting poverty and loss of status.

Community-Based Natural Resource Management

Asian countries have taken various measures in promoting community-based watershed, forest and fisheries management. Pilot experiments in community-based wildlife and protected area management have also been introduced. Continuing unwillingness of State officials to hand over authority over resources to local users, even though the most successful examples of community-based natural resource management in the region are those that have transferred rights to local communities, built on normal institutions and given scope for indigenous peoples' knowledge and traditional skills.

In the context of India, problems and experiences regarding land reformation and administration in the autonomous region are as follows:

In India, Adivasi are estimated as 7 per cent of the population of India. These 70 million people speak some 200 distinct languages and are concentrated in the "tribal belt" of Central India, with a second concentration in the North-East. These areas also contain the primary forests of the country. The traditional *zamindari pratha* was finished and introduced Land Settlements for securing the tribal residence in these areas. In 1901 the Land Revenue Code initiated to prevent the sale of tribal land without permission of the Collector. The *adivasi* are recognized as "Scheduled Tribes" and established them in "Scheduled Areas" to protect them. In the Constitution of India, the Fifth Schedule for tribes in Peninsular India and the Sixth Schedule for tribes in the North-East India also ratified in the International Labour Organization's (ILO) Convention which recognizes *adivasi* rights to the collective ownership of their traditional lands. After independence it is estimated that 6,00,000 tribal people have been removed to protected areas.

In North-East India, where formal systems of administration were extended only towards the end of British rule and then only lightly and indirectly, the Sixth Schedule recognizes Autonomous District Councils, gives them authority over land and recognizes collective land ownership.

The autonomy of East and North-East India historically, geographically and culturally are differed from each other but there are several similarities. These similarities branch partly from the contiguous geographical location and partly from a shared colonial experience.

East India's population density is among the highest in the world. The region inhabited by various tribes and ethnic groups. The administrators and politicians of India use to think that the North-East was actually seen as part of the old colonial province of Assam. The constitutional committee was faced with the aspirations of the tribal leaders of autonomy while Assamese politicians wanted more federal powers to be provided to the State. The matters related to finance and immigration. The Nga ethno-nationalist demanded that a people have the right to self-determination and that as a people. The Nagas had asserted that they sought to be free.

The Tripura Tribal Areas Autonomous District Council (TTAADC) was set up in 1979 under the Seventh Schedule to the Constitution of India. The Council came under the purview of the Sixth Schedule with effect from 1st April 1985 by fortyninth amendment to the Constitution of India. The main purpose is to protect and safeguard the tribal communities in respect of lands, forests, social customs and practices.

The Boro movement constituted for self-determination. The Boro people are in a minority. The Sanmilita Janagoshti Sangram Samiti (SJSS) that have been formed to protect the lands of other ethnic groups in the newly constituted Boro Territorial Council.

The relevance of the existing Sixth Schedule provisions for the hill areas of North-East India is doubtful. The required institutional arrangement will have to one way or another combine autonomy in regulatory functions as provided initially in the Sixth Schedule provisions and autonomy in developmental functions as provided in the Seventy-Third Constitutional Amendment.

It is important that creation of a State is an act of Parliament. Therefore, the indigenous people of the region, who scarcely contribute more than 30 members of Parliament, can hardly be considered to be "represented" as such. The whole structure of granting Statehood depends on a majority in Parliament. Therefore, Statehood politics is articulated towards convincing three hundred odd members from the most populated States in the sub-continent that the demand for an ethnic homeland.

The Fifth Schedule is meant to protect the interest of the smaller tribal population groups who are placed within the larger unit-State-structure. It provides a limited platform by way of formation of 'Tribes Advisory Council'.

The Sixth Schedule provides for autonomous District Council as well as Regional Council having powers on all the three areas of autonomy like executive, legislative and judicial. The District Council or Regional Council is empowered to make laws, for areas under its jurisdiction, relating to (i) land, (ii) forest, (iii) waterbodies, (iv) regulation of shifting cultivation, (v) village or town committees, (vi) village or town administration, (vii) appointment or succession of chiefs or headmen, (viii) inheritance of property, (ix) marriage and divorce, and (x) social customs.

The District Council is also empowered to charge and collect taxes on (i) professions, (ii) trades, (iii) callings and employment, (iv) animals, (v) vehicles, (vi) boats, (vii) entry of goods into a market for sale, (viii) tolls on passengers and goods carried in ferries, (ix) the maintenance of schools, (x) dispensaries, and (xi) roads. The degree of autonomy provided to district or regional councils in terms of decentralization of executive, legislative and judicial powers enables these institutions to protect the economic and identity interests of the tribal population groups.

The Preamble of the Constitution best reflects the vision and intentions of its founding fathers as it establishes equality, justice and liberty as the cardinal principles in regulating the society and State in India and maintaining the unity and integrity of the nation.

The Constitution ensured equality of opportunity to all

irrespective of caste, gender, religion, and one's social location. The term "political autonomy" refers to the right for a community to govern its own affairs, up to a certain level. A nation with political autonomy is only allowed control of local issues, such as education, law enforcement, justice, or community services.

The justice plays an important role to seeking the autonomy and the moment of granting autonomy which can redress the injustice. The **guarantee** was an essential element of granting autonomy and guarantee significant rights over resources and seek to control the resources for seeking the autonomy. The **custody** was an important principle to specify some measures whereby actors have agency over the autonomy granting mechanism

Conclusion

Lack of adequate infrastructure, geographical isolation and the ethnic problem create trouble in the development of the Autonomous District Council areas. The area included in the District Council is very backward and underdeveloped. It is essential to mobilize resources to promote economic status of the people residing in this area. To carry out the various schemes to improve their status of fields like economic, social, cultural, business, agriculture, etc. The rights and opportunities created for the people residing in autonomous council should be accessible for them. But they are deprived from these rights and opportunities. Time has come to think jointly the Central Government, State government and autonomous council to provide facilities for those who reside in autonomous district including all of fields present in the Constitution of India.

References

1. Srivastava, Dayawanti *et al.* (ed.) (2010). "States and Union Territories: Assam: Government". *India 2010: A Reference Annual* (54th edition.). New Delhi, India: Additional Director General, Publications Division, Ministry of Information and Broadcasting (India), Government of India.
2. Ministry of Panchayati Raj (September 8, 2009). "A Note on the Backward Regions Grant Fund Programme". National Institute of Rural Development.
3. www.wikipedia.org/wiki/autonomous.

Chapter – 12

Managing Customary Laws of Land in Nagaland: Issues, Problems and Challenges

John Sema

Institutional framework of the development process is the prime mover and account for the success of the factors of production. The developmental potentials of a State become a reality when they are put in a proper institutional context and could lead to a better quality of life of the people. Examples in support of this are in plenty, both in the developed and developing countries of the world. The very distinction which exists in the case of a developed and developing countries primarily rests on the quality of institutions which govern the productive forces. The productivity is the matter of organisational mechanism, which makes and end product different from one area to another, from one society to another and from one polity to another. It is something like a beggar sitting on the bag of gold, if potentials are not attempted within the appropriate institutional climate. The levels of development which one notices in a large country like India are not only the products of resources, capital and other inputs but they owe much to the proper societal landscape. It is human organisational ingenuity which is deeply inbuilt in its history, culture and society and which gets refined with the conscious action of man in response to the changing milieu. Nagaland is no exception to this and its development is in response to the process which rests on the societal framework.

Land is the most valuable possession of mankind and man has always been jealous of its ownership, extent of areas and unlimited rights over it with the growth of population in geometrical progression and unplanned and random exploitation of land resources with industrialisation and technological development, the balance of eco-system has been disturbed and per capita availability of land is receding. It is here that the importance to regulate land economy and land management was realised with a view to use land resources in such a manner that it would instead of putting man's survival at stake, help him to lead a happy and more comfortable life.

Land Ownership Pattern in Nagaland

The origin of the pattern of landholding and ownership system may be traced back to the history with the establishment of a village. It is customary for the Nagas to have members from different clans while establishing a village due to the practice of exogamy as well as for the purpose of defence against aggression. The leader of the group allotted the plot of land to each clan and families to construct their residential house at the time of establishment of the village.

All the land within in any of the Naga tribes may be divided into four categories:

(i) Common village land;
(ii) Khel land;
(iii) Clan land; and
(iv) Individual land.

Common Village Land

Common village land comprises of practically the land for the village council, playing ground, site for educational institutions or any other common public institution like village Council Hall is constructed, forest land, woodland, etc. The villagers rear cows, mithuns in the common land. The cultivation of land of this sort is settled by a system of general consent. Lands that are under the common ownership are not the subject matter of mortgage or sale. Each and every villager has a right to enjoy the fruits of their common land. These lands are managed by the village authority and controlled by the village authority or by the village council or elders.

Khel Land or Group Land

Khel land or group land consists of the sites for (Assembly/platform) hall for the Khel people. It is also used as a watching tower, land for the village gate, *jhum* land, meadow terrace field, forest plantation, ritual ground and especially site for the boy's dormitory (sleeping house for the male youth). Only the respective Khel members have every right to enjoy their Khel land. Khel land is the common property for the respective Khel people and so it is not the subject of mortgage or sale by an individual or groups of individuals.

Clan Land

Clan land vests with the clan members of the village in perpetuity. The ownership of the land passes from generation to generation and the land is held in common by the clan people as a whole. This land includes *jhum* land as well as terrace field, forest land, meadows to be shared by all the members of the clan according to their respective needs. Allotment is made by the clan leaders themselves after clearing the jungles on a cooperative basis.

Individual Land

All the Nagas have their individual land of various kinds irrespective of the tribe they belong to. Inside the village every Naga, on an average owns a residential site on which the house is built and normally this site lies within the area of one's own clan. An individual has also cultivable land. The head of the family possess the title of the individual land. He holds the right of possession in perpetuity. This land is cultivated by the members of the family. After the death of the head/father of the family the eldest son inherits the parental house in case of the Ao's whereas in the case of Angami's and Sema's the youngest son inherits the parental house. But overall incharge of the property especially the cultivable land is in the hands of the eldest son who takes the charge in his capacity as head of the family. It is heritable by the eldest son who succeeds to the office held by his deceased father as head of a family.

Boundary Demarcation and Settlement of Disputes

Boundaries of land are demarcated by rivulets, river ranges, hill ranges, ridges, stone pillars, trees, bamboos or natural streams, etc. Removal of boundaries is considered as an offence for which the remover is liable to fines.

Any dispute relating to boundary is referred to village council by the party whose right is in jeopardy. Such cases are disposed of by the village council and village elders in the presence of both the parties after hearing the parties according to the customary procedure. Before referring the cases to the village council an attempt is made to settle the matter by the clan elder themselves. Failing this, the matter is referred to the village council.

Loss of Title due to Migration

A Naga when he migrates to another village with the intention of settling in another village permanently loses all his right and title over the land of his original village and he is debarred from claiming the same on any future date. He is entitled to sell or gift away his land but not the joint family property. His share in the joint family property would devolve on the other members of the family.

But this rule is not applicable to a person who migrates to the urban areas for the purpose of service or business but intended to come back and settle in the village after his retirement. Such a person continues to have very close link with his village and as such he continues as the title holders of the land in his village.

Self-acquired property, whether landed or other is ordinarily alienable according to the will and pleasure of the present owner. But ancestral immovable property is ordinarily inalienable except for necessity or with the consent of male descendant under exceptional circumstances. Landed property is considered to be the main source of income for the maintenance of the entire family and as such the father cannot sell or dispose of the landed property without the consent of the grown up sons. On the death of the father, the eldest son takes the place of the father and he also does not have the jurisdiction to dispose of landed properties without reasonable grounds against the interest of the other sons.

Naga customary law is not silent regarding transmission of land from one generation to another or from one person to another. Ancestral land and inherited land are not disposed of outside the family circle or clan. But if a person is compelled to sell his inherited land due to some unavoidable circumstances, he must first consult with the members of his family and clan members and see if there is any capable person to buy and retain it within the agnatic kindred. Only when there is no such capable person within the clan, the land can be disposed of to persons outside the clan. Moreover the land should not be sold to the outsider who may be from other tribes or community. This is because of boundary demarcation done between villages by the forefathers. In case, a person sells his land to the outsider, the village council will impose fine to the defaulter for the defiant of the customary laws of the community. Hence, no person dares to sell his land to the person who is outside from his village.

Rights over Land

The right of a person over a land is in the nature of a possessory title. He enjoys exclusive and full rights of using and of disposing it. The individual holding does not come to an end at the death of the holder. The right is heritable in the society in the natural way according to the rule of inheritance. There is no custom of conditional sale or alienation of the title or right on land.

Within the Naga territory there is hardly any person who has no landed property at all. So also there is not a single piece of land without any owner. An individual does not lose his right to his allotment for unreasonably long non-use. His title is indefeasible. There is no fixed period of time for the recovery or redemption of land whether rented out or pledged.

The State of Nagaland with its 19,80,602 population as per 2011 Census against population 19,88,636 as per 2001 Census thus minus 8034 population as per the latest Census. Perhaps it may be one the record in India where the census instead of increasing it has decreased. The State has total area of 16,295 sq. km having 11 (eleven) districts and 52 development blocks to provide administration for the development activities. There are village councils mostly composed of elders with an institution of the village chiefs who are either elected or hereditary. This could be supported with examples from the Angamis. Aos, Lothas and Semas Nagas. The development and upkeeps of the village life is within the preview of the council. The chief with the advice of the elders regulates *jhuming* cultivation, exploitation of forests and use of other resources. He also channelize the human resources for productive and maintenance aspects of the village.

The State administration with its various organs like the district and block administration supplements this tribal mechanism of development. The perceptions and desire of the people through the tribal set-up are being accorded full recognition in article 371(a) of the Constitution of India wherein they have agreed to accommodate the 16 points agreement between the Government of India and the State of Nagaland stated that, "No acts or laws passed by the Union Parliament affecting the following provisions shall have legal force in Nagaland unless specifically applied to it by a majority vote of the Nagaland Legislative Assembly on the issue of religious or social practices of the Nagas, Naga customary laws and procedure and civil and criminal justice. So far as these conform decision according to Naga customary laws and the ownership and transfer of land and its resources". Although the Constitution of India provides that the customary laws and tradition of Naga people is protected and that the land ownership continue to remain in the hands of the individuals, clan, communities or village as a whole, in practice there are innumerable problems and obstacles in its implementation and it requires modification or changes in ownership pattern.

As such the government is not the owner of the land and villagers are undisputed owner of the land in the State as discussed in the customary law of the land. Thus the government does not own or possess any land except those which were handed down by the British when they left and a few areas acquired by the government are either purchased from the people or by paying land compensation in lieu of job opportunity or through donations received from villagers for specific purposes of common interest. Because of these when the government wants to undertake any project or development, first they have to acquire the land from the people. Otherwise, if the people are not willing to part their land it is not possible to execute any project or development activities without which the government cannot do anything. While negotiating and tackling all these obstacles in the way of development it wasted lots of time and energy and in the process it has finally led to imbalances in its economy and the poor quality of life of majority of people.

Thus in order to speed up development process, the government should win the confidence of the people and important legislation with regard to landholding pattern should be enacted in the State Legislative Assembly in conformity with the traditional system so as to be acceptable by the people. For the traditional tribal people land is their bread and soul which is the very survival of the people. Because most of the time when governments wants to initiate development activities landownership issue became an obstacle on way for successful implementation or execution of the scheme/projects. Hence, the government should be sincere in their effort and will to solve this protracted problem which has infested the Naga society for a long time. Therefore, the State Legislative Assembly should pass an Act or resolution in the Legislative Assembly for the general interest of the State. Also the State and civil society should educate the people and create awareness that any development activities is for their welfare and benefit of the people. And this will be possible only when they win the confidence and trust of the people. The government should also create other avenues and opportunities where people can earn their livelihood otherwise this problem will continue and the State will never be able to solve this problem.

Chapter – 13

Changing Land Ownership Patterns and Its Impact on Social Relations in Sikkim

Sohel Firdos

Historical Perspective

Like other hilly and mountainous areas, the question of land has been central to the political economy of Sikkim, because of both limited availability of cultivable land and historical factors. For many centuries, feudalism had a stranglehold over land and society. Land rights were vested primarily in the nine Kazi (feudal lords) families, with each region clearly demarcated and separated from the others. An old *bustiwala* still remembers how *chakureys* and *pakhureys* worked as very subservient tenants on private and monastery estates. They held tenancy rights in lieu of compulsory manual labour to be provided by them to the lord of the estate. This was generally provided by a member of the tenant's family.

The *adhiadars* and *kutdars* were the other categories of tillers available in the villages who paid high rents in terms of produce. The *adhiadars* were engaged to cultivate on condition that they rendered half of the produce to the primary holding while *kutdars* were engaged to cultivate on condition that they rendered a stipulated amount of crops or cash to the primary holder. The children of the *kutdars*, who lay at the very bottom of the scale, had to provide other services like cattle tending. The vagaries of nature made these tenants very vulnerable as they had no protection against crop failure and the related remissions on rent realization.

Cultivation was carried out in the name of *bustiwalas* and *mandals* (village headmen) as they were the Block-appointees of the Kazis. The *mandals* in turn rented out the arable land to the individual families of cultivators. The actual tiller was a mere tenant-cultivator. Actual holdings of land extended upto 30 acres in the case of a *mandal*, and upto 200 acres for the Kazis themselves (*cf.* Revenue Circular dated 16 October 1924). The Kazis and aristocrats assessed the revenue payable by all the people settled on the land within their jurisdiction and paid only a fixed sum to the ruling house (Kotturan, 1983).

The land under the Durbar (feudal estates) and those under the monasteries were much larger. In fact, these complex layers of intermediaries produced the judicious separation of the State and the tiller that ensured the preservation of the feudal hierarchy.

J.C. White, the first British Political Officer in Sikkim, introduced a modified lease system of land tenure in 1888. This led to the emergence of new land settlement pattern that ultimately brought about a change in land ownership pattern. It resulted in the emergence of some early Nepali settlers (particularly the Newars) as lessee landlords who were later designated as *thikadars* (Sinha, 1975; Datta, 1994).

The emergence of lessee landlordism may be taken as an important turning point in ethnic relations in Sikkim. Since all registrations of land transactions were accomplished through the lessee landlords, they could absorb as many newcomers as they wanted. It was in fact easy to extract initial payments in the form of *nazaranas* (Basnet, 1974). The tenants, however, did not have any security of tenancy rights on the land under the jurisdiction of lessee landlords. Thus, they remained at the mercy of the lessees.

Ethnic Groups and Land

Land distribution and land revenue contribution according to the three major ethnic groups in the State, namely Bhutias, Lepchas and Nepalis, is highly uneven. This may be explained by historical factors. There have been several safeguards for ensuring land rights of the two indigenous tribal communities, the Bhutias and the Lepchas. The first step in this direction goes back to 1917, when the Government of Sikkim issued a notification termed as Revenue Order No. 1, which prohibited sale or transfer of land belonging to Bhutias or Lepchas to non-Bhutias or Lepchas without the permission of the State. The Notification No 3082/L.R., dated 24 March 1954, issued by the Land Revenue Department, of the Sikkim State and signed by Tashi Namgyal, the Maharaja of Sikkim, reinforced the Revenue Order No. 1 of 1917. This notification remained in force even after 1975, and has been strictly implemented.

In the late 1980s, the Sikkim Alienation of Land (Regulation) Bill 1989, and the Sikkim Transfer of Land (Regulation Bill 1989) were also passed by the State Legislature. These Bills respectively aimed at restricting alienation of land by the members of Bhutia and Lepcha communities of Sikkimese origin to persons other than Bhutia and Lepcha of Sikkimese origin and also by Sikkimese in favour of non-Sikkimese. These bills have been reserved by the Governor pending the assent of the President.

It is important to note that this regulation is applied even in cases that could conceivably be in the larger interest of the State in terms of providing employment opportunities, as in the case of private industrial enterprises. It is possible to obtain on lease land belonging to tribals, but only after obtaining the permission of the State Government.

In 1983, the Nepalis owned almost 59 per cent of the total cultivated land and contributed almost 64 per cent of the total land revenue generated in the State. In contrast, the land share of the Bhutias and the Lepchas was 20 per cent each and their land revenue contribution was 19 per cent and 16 per cent respectively. It was only with respect to land under cardamom cultivation that Nepalis did not rank first either in terms of owned areas or land revenue. Of 21,762 hectares, devoted to cardamom cultivation 33 per cent was owned by the Lepchas, followed by Bhutias (27%) and the Nepalis (22%), with the remaining 19 per cent being outside the public domain.

In fact, Sikkim has a tradition of protecting traditional land tenures and there is considerable sensitivity regarding the maintenance of customary laws governing ownership and proscribing alienation to outsiders. Some of the old laws of Sikkim have been upheld by the highest court of law in the country.

Inequitable Landholdings

Distribution of operational landholdings in Sikkim is skewed. In 1990-91 the lowest echelon of landholders, the marginal holders, representing about 50 per cent of landholdings, held 10.3 per cent of the total operational land area. The East district had the highest concentration of landholding by marginal farmers both in terms of number of holdings (56%) and area (12.8%). In contrast, farmers with large holdings (more than 10 hectares) accounted for 2.3 per cent of the landholdings, but owned 20.2 per cent of the operational area. The North district had the highest number of large farmers who, with 5 per cent of the operational holdings, owned over 30 per cent of the operated land within the district. However, except in the East district, land holdings and operated area are well distributed at least among the small and semi-medium farmers.

Inequality also characterizes land distribution among farmers belonging to Scheduled Castes (SCs). On one hand, large farmers, who are not even 1 per cent (0.4%) of the total number of SC farmers, held nearly 10 per cent of the operated area. On the other, marginal farmers, representing 70 per cent of SC farmers, operated 27.2 per cent of the operated area. A positive sign, however, is the fact that small farmers, representing over 19 per cent of the operated holdings, owned almost 28 per cent of the operated area.

In the case of the Scheduled Tribe (ST) farmers, land distribution is not as skewed. The number of marginal farmers is much lower (42%) whereas the semi-medium and medium farmers together owned more than 35 per cent of the operated holdings and over 57 per cent of the operated land. However, in this case also, large farmers, who constituted hardly 5 per cent of the ST farmers, had a giant share of over 28 per cent of the land area under their possession. This is indicative of the fact that the traditional patterns of land holdings in Sikkim have not undergone any significant change.

In Sikkim, as elsewhere in the country, the movement for democracy and greater political and civil rights, was based on the demand for the abolition of the *zamindari system*[1]. The abolition of the *Zamindari* system in 1949 was immediately followed by an official notification making it compulsory for all revenues raised from land by revenue agents to be deposited directly with the government. However, the private estates of the *Durbar* (feudal estates), and the monastery land were left untouched by this new land regime. As the debate on land reforms and its vital role in initiating changes in the agrarian society for a rapid transformation picked up, land reform issues gradually gained ground, with both political and social dimensions.

Until 1976, there were only four District Collectorates. The District administration was headed by the District Collectors, reporting to the Secretary, Land Revenue Department, whose office was responsible for maintaining the records of rights. Accordingly, the Directorate of Survey and Settlement (now known as New Land Record Office) was established in 1976, to ascertain the records of rights and the ownership of land through periodic survey operations.

The first scientific survey of land in Sikkim was carried out in 1950-58, using the British measurement system of acres and miles. The second, 1976-83 survey is till date the last land survey carried out after Sikkim became a constituent State of India. The metric system of hectares and kilometres was used, and the survey covered all the areas of Sikkim. This was a critical survey, because previous land records had undergone several changes due to partition, mutation, registration and acquisition by the government, private parties and others.

In the 1976-83 round of survey, the age-old *elakas* were delimited and revenue blocks similar to those in other States were created. At present, there are 411 such revenue blocks, which are divided among 40 village level circles comprising 10 to 11 revenue blocks per circle.

The next step in the improvement and modernization of land records was their computerization, which was completed in 1998.

Computerization of Land Records

Sikkim has the rare distinction of being the first State in the country to have fully computerized land records. The work started in 1992, and all the land records as on January 1, 1998 have been documented. They include records of revenue administration and individual rights.

Each landholder can now have a computerized record of his/her land rights. As computers have been installed in both the district and sub-division offices, the records can be updated on a regular basis. This can even provide online mutation.

The following features characterize land relations in Sikkim:

(i) *Small Operational Holdings and Declining Cultivation*: The total land under operational holding in 1991, was hardly 16 per cent, while the remaining 84 per cent was taken up by trees and groves (42%), barren land (22%) and permanent pastures and grazing land (17%). However, after the merger with the Union of India and the consequent introduction of constitutional democracy; there has been a gradual increase in the total land area for utilization by over 32,000 hectares during the fifteen-year period between 1976-77 and 1991-92. This is mostly due to the increase in cultivable wasteland and land available for cultivation. In spite of this increase, the net area sown has gone down between 1980-81 and 1990-91 (Table 13.1).

Table 13.1: Land-Use Pattern in 1981 and 1991

Classifications	*1980-81*		*1990-91*	
	Hectares	*(%)**	*Hectares*	*(%)**
Net Area Sown	78381	11.04	63254	8.91
Area under current fallow	4428	0.62	3906	0.55
Other uncultivated area excluding fallow land	4560	0.64	10830	1.53
Fallow other than current fallow	9474	1.34	9204	1.30
Cultivable wasteland	681	0.10	9807	1.38
Land not available for cultivation	11604	1.64	14300	2.02
Total operated land	109068	15.37	111302	15.69

Note: * Indicates the percentage distribution of different land-use heads. The total geographical area is 709,600 hectares.

Source: Reproduced from Sikkim Human Development Report, 2001. This report cites its source as *Government of Sikkim (1991), State Report on Agriculture—Census, 1990-91, Bureau of Economics and Statistics, Gangtok.*

(ii) *Declining Per Capita Availability:* Per capita availability of net cultivated area has also recorded a sharp decline from

0.31 hectare in 1971 to 0.27 hectare in 1981 and 0.17 hectare in 1991. However, wide variations continue to exist across districts. The extent of fragmentation and partition of land can be assessed from the phenomenal increase in the number of land holdings from 17,000 in 1950-58 to 54,500 as per the survey of 1976-83.

An analysis of the district-wise distribution of cultivated land shows that in 1976-83 the share of West district was the highest with 29.34 per cent of the total cultivated area, followed by East (28.96%) and South (26.95%). The North district with a meagre share of 14.76 per cent has the lowest cultivated land availability. With respect to land revenue, however, the East district contributed the largest share (33.64%) of the total revenue collection (Table 13.2).

Table 13.2: District-wise Land Distribution and Revenue in Sikkim (1976-83)

Districts	*Area*		*Land revenue*	
	Hectares	*% share*	*Rs.*	*% share*
North	16228.00	14.75	24761.51	8.75
East	31846.16	28.96	95192.29	33.64
South	29629.77	26.95	81962.99	28.96
West	32259.07	29.34	81079.62	28.65
Total	109963.00	100.00	282996.40	100.00

Source: Reproduced from Sikkim Human Development Report, 2001. *This Report cites the source as Government of Sikkim, Land Records Section, Department of Land Revenue, Gangtok.*

Out of the total cultivated land of 109,963 hectares, dry land constituted 58 per cent, followed by Cardamom (19%), paddy (13%) and wasteland (10%). This reflects the cropping pattern in Sikkim where different districts were prominent for each of the above four categories. The maximum dry land area was in the West district (35%), followed by the South district (33%), the East district (21%) and the North district (11%). The largest portion of cardamom land was in North (32%), followed by East (29%), South (21%) and West (18%). In case of paddy, the South district ranked first (44%), followed by West (25%), East (23%) and North (8%). It may be noted that cardamom, the most vital cash crop, is predominantly grown in the tribal belt in the North district.

Land Tax and Revenue

Levy on land used to be the major source of government revenue. In 1889, land revenue constituted as high as 66 per cent of the total revenue of Sikkim. The rest came from forest (13.5%), excise (7.1%) and from miscellaneous activities. In 1930-31 land revenue realization used to be 27.5 per cent of the total revenue mobilization (Debnath, 1974). However, in the post-merger period, the contribution of land revenue to the State exchequer has been very small both because of low rates imposed by the State and the sensitivity involved in dealing with land issues. This pattern has persisted in spite of the fact that revised land rent rates were announced in 1998, and more land was diverted from the traditional system of subsistence farming to commercial cultivation.

Land Reforms

After changes in land regulations were introduced in 1949, the State of Sikkim issued a Notification No. 3082/L.R., in 1954, which has some progressive elements of land reforms and provides for the maintenance of economic holdings. The lower ceiling of landholding is ensured by the provision of the 'sale of land in execution'. It is stated that:

> *(....) No Court will sell or transfer a holding or any part of a holding of a primary holder in execution of a decree, whether revenue or civil, if by such sale or transfer the said holding will become less than five acres in area. provided that sale of land to meet Government dues will be excluded from the purview of this notification.*

This notification also fixed an upper ceiling through 'restrictions on purchase', according to which 'No person who already has a holding or holdings exceeding 20 acres in area may purchase land sold in execution of a revenue or civil decree'.

In a mountainous State like Sikkim, the all India land ceiling of 12 acres, which is a blanket ceiling irrespective of topographical variations, is difficult to implement. It might be more appropriate to fix land ceilings, on a case to case basis using methods suitable to the terrain.

After the merger of Sikkim in 1975, the government intervened mainly to provide legislative measures against the termination of cultivation rights and for the continuity of cultivation by existing cultivators. This was done in view of the problems faced by the tillers of the soil, who cultivated land owned by others under precarious terms and conditions. Among the numerous land reform measures adopted by the State Government; the most notable are the enactments of Sikkim Agricultural Land Ceiling and Land Reforms Act 1977, Cultivators Protection Act 1985, and the Land Bank Scheme of 1995. These acts and measures helped to make the transition from a feudal past to a more egalitarian society, relatively smooth.

The Land Bank Scheme is the latest welfare scheme introduced by the Government of Sikkim to consolidate land reform measures in the State. Under this scheme the landowners donate a certain part of their land, entirely voluntarily, to the government. This land is then distributed to the landless (*Sukumbasis*), identified by the department concerned on the basis of a detailed survey. A landless household is now entitled to half an acre of land, with the settlement deed carrying the names of both husband and wife. The upper limit of the land cost has been fixed at Rs 25,000.

It is important to note that this Scheme puts the onus of looking for suitable cultivable land on the beneficiary who can select the land of his choice. Over 301 families have benefited from this Scheme. The success of this scheme reinforces the fact that the programme of land reforms implemented so far has not led to any significant redistribution of land in Sikkim. This has had adverse effects on both social cohesiveness and agricultural productivity.

Note

1. This is a system of land tenure in which the principal landowner or Zamindar is a rent receiver rather than a cultivator. Its beginning in modern India goes back to the Permanent Settlement of 1793.

Index